FIBER OPTICS

Editor

James C. Daly
Professor of Electrical Engineering
University of Rhode Island
Kingston, Rhode Island

CRC Press, Inc.
Boca Raton, Florida

06063206

Library of Congress Cataloging in Publication Data

Main entry under title:

Fiber optics

Bibliography: p.
Includes index.
1. Fiber optics. I. Daly, James C., 1938-
TA1800.F516   1984   621.36'92   83-15266
ISBN-0-8493-5103-0

D
621.3692
F1B

Direct all inquiries to CRC Press, Inc., 2000 Corporate Blvd., N.W., Boca Raton, Florida, 33431.

© 1984 by CRC Press, Inc.
Second Printing, 1985
Third Printing, 1985

International Standard Book Number 0-8493-5103-0

Library of Congress Card Number 83-15266
Printed in the United States

PREFACE

This book is designed to provide communication engineers with the theoretical and practical knowledge needed to understand and design fiber optic communication systems. Since this is a new technology, the University study of most practicing engineers did not include courses on fiber optic communications. These people do, however, have an understanding of communication systems and the fundamentals of electromagnetic theory. This book bridges the gap between classical communication practice and the new techniques required to design fiber optic communication systems.

The book is organized into eight chapters, each written by an expert with first-hand experience. Chapter 1 reviews general properties of fibers, history, and some applications. This includes a review of the electromagnetic base of some important optical phenomena, matrix methods for ray tracing, and Gaussian beams. Chapter 2 covers optical waveguide manufacture. Various types of fibers and the processes used to produce them are explained. This includes preform preparation and fiber drawing. Measurements used to characterize optical and mechanical properties of fibers are covered. Chapter 3 explains propagation in fibers in terms of both wave and ray theory. Dispersion, coupling, modal noise, and the influence of fiber properties on system performance are covered. Sources are discussed in Chapter 4. Characteristics pertinent to optical communication systems, including physical principles and structures of lasers and IRED diodes are explained. Chapter 5 covers detectors. The principles of operation of Si PIN and avalanche photo diodes are discussed. Detectors for long wavelength systems and noise in detectors is covered. Chapters 6 and 7 treat fiber optic communication systems. Chapter 6 covers the important topic of digital communication systems. The operation of various components, such as fibers and lasers, are reviewed in terms relevant to communication engineers. This is followed by an analysis of design criteria for digital receivers. Practical constraints are covered. Analog systems are covered in Chapter 7. This includes a comprehensive treatment of the fundamentals and limitations of analog transmission systems. Noise models and degeneration effects are covered. Practical considerations in implementing systems are included. Chapter 8 treats the use of optical fibers in imaging systems. Imaging theory, fiber fabrication, and imaging applications are discussed.

The editor acknowledges the advice and assistance of Billy Burdine and Dr. John F. Ambrose of GTE Laboratories and Fred Allard of the Naval Underwater Systems Center.

THE EDITOR

James C. Daly, Ph.D., is Professor of Electrical Engineering at the University of Rhode Island, Kingston, Rhode Island.

Dr. Daly received his B.S. degree in Electrical Engineering from the University of Connecticut in 1960 and his M.E.E. and Ph.D. degrees from Rensselaer Polytechnic Institute in 1962 and 1967, respectively.

Dr. Daly is a member of the Institute of Electrical and Electronics Engineers and the honor societies, Tau Beta Pi, Eta Kappa Nu, and Sigma Xi.

His current research interest involves VLSI for fiber optic electronics. He has published over twenty papers in the area of optical communications and electronic systems. He holds a patent on an optical waveguide.

CONTRIBUTORS

Ishwar D. Aggarwal
Vice President
Research and Development
Valtec
West Boylston, Massachusetts

E. E. Basch
Program Manager
Optical Communication Systems
GTE Laboratories
Waltham, Massachusetts

Howard A. Carnes
Senior Member
Technical Staff
GTE Laboratories
Waltham, Massachusetts

James C. Daly
Professor of Electrical Engineering
University of Rhode Island
Kingston, Rhode Island

To R. Hsing
Senior Member
Technical Staff
GTE Laboratories
Waltham, Massachusetts

T. P. Lee
Member, Technical Staff
Crawford Hill Laboratory
AT&T Bell Laboratories
Holmdel, New Jersey

Gabriel Lengyel
Professor of Electrical Engineering
University of Rhode Island
Kingston, Rhode Island

Robert G. Olsen
Professor of Electrical Engineering
Washington State University
Pullman, Washington

David A. Rogers
Associate Professor of Electrical
Engineering
North Dakota State University
Fargo, North Dakota

TABLE OF CONTENTS

Chapter 1

INTRODUCTION

James C. Daly

TABLE OF CONTENTS

I. HISTORY

The use of fibers for optical communications was suggested by Kao and Davies in 1968.[1] At that time typical fiber losses were above 1000 dB/km. Kao suggested that purer materials should permit much lower losses. In November of 1970, Kapron et al. reported an observed total attenuation of approximately 20 dB/km in a single mode fiber.[1] Today, fibers operate at a wavelength of 1.5 μm with losses less than 1 dB/km.[3-4] Low losses have been achieved by reducing impurity absorption due to transition metal ions such as iron, chromium, cobalt, and copper. Absorption from OH^- ions due to water impurity is also an important factor. Parts per billion purity of iron and chromium ions is required if their loss contributions are to be kept below 1 dB/km.[5]

A communcation system requires a transmitter, a transmission medium, and a receiver. Fiber optics became a feasible transmission medium in 1970 with the reduction of losses to 20 dB/km. At that time, technology also existed to produce semiconductor detectors suitable for use with optical fibers, but there were no suitable sources. The earlier invention of the laser and the possibility of using it for communication had stimulated fiber optics research. In 1970 laser sources and light emitting diodes (LEDs) had problems of short lifetime and low outputs. Early semiconductor lasers were inefficient and required cooling. A series of improvements has resulted in more efficient semiconductor lasers operating reliably at room temperature with high outputs and lifetimes greater than 10^5 hr. Today, with high quality semiconductor sources, low-loss fiber transmision media and low noise semiconductor detectors, the three elements (transmitter, medium, and receiver) are available for the construction of economical, reliable optical communication systems.

A. Fiber Classes

The information-carrying capacity (bits/sec) of an optical fiber is determined by its impulse response. The impulse response and thus the bandwidth are largely determined by the modal properties of the fiber. The optical fibers, in common use, can be separated into two classes based on their modal properties, (1) single-mode fibers, and (2) multimode fibers. Single-mode fibers are step-index. Multimode fibers can be divided into step- and graded-index. Step- or graded-index refers to the variation of the index of refraction with radial distance from the fiber axis. Fiber types are discussed thoroughly in Chapter 2. Figure 1 of Chapter 2 shows these three types of fibers (1) step-index multimode, (2) graded-index, and (3) single-mode. These fibers consist of a core surrounded by a cladding. The higher index of refraction of the core compared to the cladding causes total internal reflection at the core cladding interface in step-index fibers. In graded-index fibers, the gradual decrease in the index of refraction with distance from the fiber axis causes light rays to bend back toward the axis as they propagate. Multimode guides are characterized by multiple propagation paths for rays. A modal description of multimode fibers shows different propagation velocities for different modes. Therefore, energy input into the fiber from a short pulse, coupled into a multiple of modes, will arrive at the receiving end of the fiber distributed over a time interval. The spreading out in time of the received pulse is due to the differing propagation delays of the different modes. This pulse spreading is referred to as modal dispersion. It reduces the information capacity of the fiber by limiting the number of distinct pulses that can be transmitted in a given time interval. Graded-index guides have less modal dispersion than step-index guides. Of course, modal dispersion does not occur in single-mode guides where only one mode propagates. Pulse broadening in single-mode fibers is due to material dispersion and the dispersion associated with the waveguide mode. Single-mode fibers can be designed so that these two sources of dispersion cancel at a particular wavelength.

B. Applications

Optical fibers have advantages that make them attractive in a variety of applications. They have extremely high bandwidth. Their small diameter and high tensile strength result in smaller, lighter weight cables and connectors. Since they are electrical insulators, optical fibers are immune to inductive interference and are not subject to ground loop problems. They can be used in high voltage environments without providing unwanted conduction paths. They do not radiate electromagnetic energy. In addition, they are tolerant to temperature extremes, resist corrosion, are reliable, and easily maintained. The raw material used to fabricate glass fibers is sand, an abundant resource.

First generation optical fiber communication systems operate at a wavelength of 0.82 μm. Second generation systems, operating at a longer wavelength of 1.3 μm where fiber losses and material dispersion are less, offer significantly less attenuation and greater bandwidth.

Optical fiber applications have been pioneered by telephone companies. General Telephone of California installed the first optical fiber link carrying regular telephone service in Long Beach, Calif., on April 22, 1977.[6] It was a 1.544 Mb/s link utilizing a graded-index fiber with a 6.2 dB/km mean loss, LED sources, and avalanche photodiode detectors. Only two repeaters were used with the 9.1 km link. An equivalent metallic link would have required five repeaters for the same data rate.[5] The current Bell System fiber optic program centers around digital trunk transmission at 44.7 Mb/s.[7] As many as 672 voice circuits are transmitted in a pair of fibers. Bell is planning a transatlantic cable operating at 1.3 μm. Each section of cable uses a laser and one or more standbys to achieve a mean time between failures of 8 years. Repeaters are at about 30 km intervals. Multiple single-mode fibers capable of carrying more than 4000 voice circuits are planned.[7]

CATV systems use fibers to transmit signals from earth stations to studio facilities.[9] Distribution of CATV signals from studios to subscribers is being tested.[10]

Optical fibers have promise for use in computer systems.[11] It is attractive to replace parallel interconnects with serial fiber optic links. Cable and connector bulk is significantly reduced. Reliability is improved. Fiber optic high data rates, noise immunity, and low loss make it possible to extend high data rate channel links beyond the confines of the computer room. The use of smart terminals increases the need for high bandwidth local networks. Smart terminals process and store information. Relatively large amounts of information are transmitted to the host computer in bursts. A large bandwidth is required if excessive response delays are to be avoided.[11] The interest of the military has stimulated a wide range of fiber applications, from rotation and sonar sensors to communication links. Fibers permit dramatic weight and bulk reduction and provide large bandwidth and high reliability. A 64-km fiber optic field link, used by the Army, transmitting 2.3 Mb/s requires seven repeaters and can be transported on one $2^{1}/_{2}$ ton truck. The equivalent coaxial link requires 39 repeaters and four $2^{1}/_{2}$ ton trucks for transportation.[12] Fiber optic sonar links have been developed to transmit information from external sensors through the submarine hull to inboard signal processors.[13] These links reduce the size of submarine hull penetrators in addition to improving system performance.

One of the original applications of fibers is image transmission. The flexible fiberscope has been widely used in medicine since the 1950s. Modern fiber optics technology has recently been applied to office copy machines. This is discussed further in Chapter 7.

The following three sections of this chapter discuss basic electromagnetic theory and optics useful in describing phenomena in fiber optic systems. Section II reviews electromagnetic plane waves. Section III illustrates some examples of matrix methods for the description of ray propagation. Section IV covers some properties of optical beams.

II. WAVE OPTICS

The electromagnetic description of optics forms a basis for the explanation of a number of phenomena occurring in optical fiber systems. Reflection at a dielectric interface occurs when light passes through a lens, is injected into a fiber, or is reflected at the fiber core cladding interface. Phenomena such as total internal reflection, penetration of the optical field into the cladding, and Snell's law follow from the description of the reflection of plane waves from a dielectric interface.

A. Maxwell's Equations

Maxwell's equations written in differential form are

$$\vec{\nabla} \times \vec{E} = -\frac{\partial \vec{B}}{\partial t} \tag{a}$$

$$\vec{\nabla} \times \vec{H} = J + \frac{\partial \vec{D}}{\partial t} \tag{b}$$

$$\vec{\nabla} \cdot \vec{D} = \rho \tag{c}$$

$$\vec{\nabla} \cdot \vec{B} = 0 \tag{d}$$

$$\tag{1}$$

where E,B,H,J,D and ρ are the electric field intensity, the magnetic flux density, the magnetic field intensity, the current density, the electric flux density, and the electric charge density, respectively. The del operator is ∇. In rectangular coordinates $\nabla = x\,\hat{x} + y\,\hat{y} + z\,\hat{z}$, where $\hat{x}$, $\hat{y}$, and $\hat{z}$ are unit vectors in the x, y, and z directions.

In a source-free region such as air or glass with no free charge, $\vec{J} = 0$ and $\rho = 0$. Also, if the medium is time invariant, homogeneous, isotropic, and linear, then $\vec{B} = \mu\vec{H}$ and $\vec{D} = \epsilon E$, where the permeability, μ, and the permittivity, ϵ, are scalar quantities that do not vary in space or time. Under these conditions Maxwell's equations are

$$\vec{\nabla} \times \vec{E} = -\mu\frac{\partial \vec{H}}{\partial t} \tag{a}$$

$$\vec{\nabla} \times \vec{H} = \epsilon\frac{\partial \vec{E}}{\partial t} \tag{b}$$

$$\vec{\nabla} \cdot \vec{E} = 0 \tag{c}$$

$$\vec{\nabla} \cdot \vec{H} = 0 \tag{d}$$

$$\tag{2}$$

The wave equation is obtained by combining Equation 2a and 2b. The first step is to take the curl of Equation 2a.

$$\vec{\nabla} \times (\vec{\nabla} \times \vec{E}) = \vec{\nabla} \times \left(-\mu\frac{\partial \vec{H}}{\partial t}\right) \tag{3}$$

Interchanging the order of differentiation with respect to space and time, and moving the constant μ out of the derivative

$$\vec{\nabla} \times (\nabla \times \vec{E}) = -\mu\frac{\partial}{\partial t}(\vec{\nabla} \times \vec{H}) \tag{4}$$

Substituting Equation 2b for $\vec{\nabla \times H}$,

$$\vec{\nabla \times (\nabla \times E)} = -\mu\epsilon \frac{\vec{\partial^2 E}}{\partial t^2} \tag{5}$$

Recall the vector identity $\vec{\nabla \times \nabla \times E} = \vec{\nabla (\nabla \cdot E)} - \vec{\nabla^2 E}$. Applying this vector identity to Equations 5 and using Equation 2c, results in the vector wave equation,

$$\vec{\nabla^2 E} = \mu\epsilon \frac{\vec{\partial^2 E}}{\partial t^2} \tag{6}$$

When the field vectors have sinusoidally varying components it is convenient to use phasors to represent the components. Consider the electric field,

$$\vec{E} = \hat{x}\, A\, \cos(\omega t + \theta_x) + \hat{y}\, B\, \cos(\omega t + \theta_y) + \hat{z}\, C\, \cos(\omega t + \theta_z) \tag{7}$$

The x, y, and z components vary sinusoidally at the same frequency. Each component may have a different amplitude and phase. When phasors are used to represent the sinusoidally varying components the following phasor vector results

$$\vec{E} = E_x\, \hat{x} + E_y\, \hat{y}\, E_z\, \hat{z} \tag{8}$$

where the complex numbers $E_x = A <\Theta_x$, $E_y = B <\Theta_y$, and $E_z = C <\Theta_z$ are the phasor representations of the components of the vector.

Taking the derivative of a sinusoid with respect to time corresponds to multiplying its phasor representation by $j\omega$. Therefore, when field quantities are represented by phasors the wave equation is

$$\vec{\nabla^2 E} = -\omega^2\mu\epsilon \frac{\vec{\partial^2 E}}{\partial t^2} \tag{9}$$

B. Plane Waves

Waves with planar constant phase surfaces are called plane waves. Although plane waves have a simple mathematical form, their behavior is of general interest. Complex waves can be represented as a sum of plane waves, and in the neighborhood of a point all waves appear to be plane waves. When the amplitude and phase of field quantities are constant on a plane, the wave is called a uniform plane wave. A simple solution to Equation 9 is a uniform plane wave, polarized in the x direction and propagating in the z direction. A wave polarized in the x direction has electric field components only in the x direction. Assume the phasor vector representation of the electric field is only a function of z

$$\vec{E} = \hat{x}\, E_x\,(z) \tag{10}$$

Since for this case there is no dependence on x and y, the Laplacian operator reduces to, $\nabla^2 = \partial^2/\partial z^2$. The wave equation becomes

$$\frac{\partial^2 E_x}{\partial z^2} = \omega^2\mu\epsilon E_x \tag{11}$$

The solution to Equation 11 is

$$E_x = E_o e^{\pm jkz} \tag{12}$$

where $k = \omega\sqrt{\mu\epsilon}$. $E_o = A < \Theta$ is a complex constant depending on the boundary conditions. E_x given by Equation 12 is a phasor quantity that corresponds to the following time function

$$E_x (t,z) = A \cos (\omega t \pm kz + \theta) \tag{13}$$

A point of constant phase for the wave given by Equation 13 is

$$\text{const} = \omega t \pm kz \tag{14}$$

The phase velocity v'_p, is obtained by differentiating Equation 14 with respect to time

$$0 = \omega \pm k \frac{dz}{dt} \tag{15}$$

The velocity of a constant phase point is

$$v_p = \frac{dz}{dt} = \pm \omega/k = \pm \frac{1}{\sqrt{\mu\epsilon}} \tag{16}$$

A plus sign before k in Equation 16 corresponds to a wave propagating in the negative z direction, and a minus sign corresponds to a wave propagating in the positive z direction. One wavelength is the distance for a 2π phase shift. Therefore, $2\pi = k\lambda$ or $k = 2\pi/\lambda$ where λ is the wavelength.

The index of refraction of a medium is defined as the ratio of the phase velocity in free space to the phase velocity in the medium. It follows from Equation 16 that the index of refraction is

$$n = \frac{v_p}{v_p} = \frac{k}{k_o} = \sqrt{\frac{\epsilon}{\epsilon_o}} = \sqrt{\epsilon_r} \tag{17}$$

where c is the velocity of light in free space, $k_o = \omega \sqrt{\mu_o\epsilon_o}$, ϵ is the permittivity of free space, and ϵ_r is the dielectric constant.

1. Generalized Plane Waves

Constant phase surfaces for the plane waves given by Equation 12 are planes perpendicular to the z axis ($x =$ constant). Consider the same wave propagating in some direction other than along the z axis. Such a wave would result if the coordinate system were rotated. The constant phase surface would be a plane described by $k_x\hat{x} + k_y\hat{y} + k_z\hat{z} = \phi =$ constants. A field quantity associated with this wave is

$$E = E_o e^{-j(k_xx + k_yy + k_zz)} \tag{18}$$

where E_o is a constant phasor vector. Recall $\vec{r}$, the vector position of any point in space, $\vec{r} = x\hat{x} + y\hat{y} + z\hat{z}$. Also define the propagation vector $k = \vec{k}_x\hat{x} + k_y\hat{y} + k_z\hat{z}$. The constant phase point, ϕ, may be written $\phi = \vec{k} \cdot \vec{r} = k_xx + k_yy + k_zz$

$$E = E_o e^{-jk \cdot r} \qquad (19)$$

Taking derivatives with respect to x of plane wave field quantities with exponential spatial variation such as given in Equation 18 is equivalent to multiplying by $-jk_x$. Derivatives with respect to other coordinates may be treated similarly. This allows the del operator, which in rectangular coordinates is $\nabla = \hat{x}\, \partial/\partial x + \hat{y}\, \partial/\partial y + \hat{z}\, \partial/\partial z$, to be replaced by, $-jk$. Also, the Laplacian operator which in rectangular coordinates is $\nabla^2 = \partial^2/\partial x^2 + \partial^2/\partial y^2 + \partial^2/\partial z^2$, is replaced by

$$\nabla^2 = -k_x^2 - k_y^2 - k_z^2 = -k^2 \qquad (20)$$

where k^2 is the magnitude of $\vec{k}$ squared.

Substituting Equation 19 into Maxwell's equations, (Equations 2) results in the following representation of Maxwell's equations for plane waves.

$$\vec{k} \times \vec{E} = \overrightarrow{\mu\omega H} \qquad \text{(a)} \qquad (21)$$

$$\vec{k} \times \vec{H} = -\epsilon\omega \vec{E} \qquad \text{(b)}$$

$$\vec{k} \cdot \vec{E} = 0 \qquad \text{(c)}$$

$$\vec{k} \cdot \vec{H} = 0 \qquad \text{(d)}$$

The magnitude of k is found by applying Equation 20 to Equation 9:

$$k = \omega\sqrt{\mu\epsilon} \qquad (22)$$

The magnetic field associated with the plane wave is obtained using Equation 21a

$$\vec{H} = \frac{1}{\omega\mu} \vec{k} \times \vec{E} \qquad (23)$$

It follows from Equation 21b that $\vec{E}$ is perpendicular to both $\vec{k}$ and $\vec{H}$. Since $\vec{k}$ and $\vec{E}$ are perpendicular, the magnitude of $\vec{H}$ is the product of the magnitudes of $\vec{k}$ and $\vec{E}$ divided by $\omega\mu$. Using Equations 23 and 22 it follows that

$$E = \sqrt{\frac{\mu}{\epsilon}}\, H \qquad (24)$$

where E and H are the magnitudes of $\vec{E}$ and $\vec{H}$ and $\sqrt{\mu/\epsilon}$ is the characteristic impedance of the medium. Note that

$$Z = \sqrt{\frac{\mu}{\epsilon}} = \frac{Z_o}{n} \qquad (25)$$

where n is the index of refraction and Z_o is the characteristic impedance of free space.

It follows from Equations 21a and 21b that $\vec{E}$, $\vec{H}$, and $\vec{k}$ form an orthogonal set of vectors. $\vec{E} \times \vec{H}$ is in the direction of $\vec{k}$. Since $\vec{E}$ and $\vec{H}$ are perpendicular, the magnitude of $\vec{E} \times \vec{H}$ is the magnitude of $\vec{E}$ multiplied by the magnitude of $\vec{H}$. From this and Equation 24 it follows that

$$\overrightarrow{E} \times \overrightarrow{H} = \frac{E^2}{Z} \hat{k} \tag{26}$$

where $\hat{k}$ is the unit vector in the direction of $\overrightarrow{k}$. $\overrightarrow{E} \times \overrightarrow{H}$ is the Poynting vector. Its units are watts per square meter (W/m²). The Poynting vector is in the direction of energy flow in isotropic media.

C. Reflection at a Dielectric Interface

The amount of light reflected at a dielectric interface depends on the polarization, the angle of incidence, and the indices of refraction of the two media. The reflection and transmission coefficients are found by satisfying the boundary conditions. At the boundary conditions, the tangential components of both the E and H vectors are continuous across the interface.

A representation of a plane wave reflection is shown in Figure 1. The plane wave shown is polarized parallel to the plane of incidence. The plane of incidence is defined as the plane containing the normal to the surface and the k vector of the incident plane wave.

The incident, reflected and transmitted plane waves are

$$\overrightarrow{\mathscr{E}}_i = \overrightarrow{E}_i \cdot e^{-j\overrightarrow{k_i} \cdot \overrightarrow{r}}$$

$$\overrightarrow{\mathscr{E}}_R = \overrightarrow{E}_R \, e^{-j\overrightarrow{k_R} \cdot \overrightarrow{r}}$$

$$\overrightarrow{\mathscr{E}}_t = \overrightarrow{E}_t \, e^{-j\overrightarrow{k_t} \cdot \overrightarrow{r}} \tag{27}$$

1. Snell's Law

Snell's law follows from the boundary conditions and states that the phase variation along the interface must be the same for the incident, reflected, and transmitted fields. That is, at the interface

$$\overrightarrow{k}_i \cdot \overrightarrow{r} = \overrightarrow{k}_R \cdot \overrightarrow{r} = \overrightarrow{k}_t \cdot \overrightarrow{r} \tag{28}$$

At the interface $z = 0$. Also in Figure 1, the coordinate system has been drawn with the incident k vector in the xz plane; therefore, the y component of $\overrightarrow{k}_i = 0$. In this situation, the phase variation along the interface depends only on x. Equating the phases of the incident, reflected, and transmitted waves at the interface.

$$n_1 k_o \sin \theta_i = n_1 k_o \sin \theta_R = n_2 k_o \sin \theta_t \tag{29}$$

where $n_1 k_o \sin \Theta_i$, $n_1 k_o \sin \Theta_R$, and $n_2 k_o \sin \Theta_t$ are the x components of the incident, reflected, and transmitted propagation vecors, respectively.

Equation 29 leads to two important observations. The first is that the angle of incidence equals the angle of reflection, $\Theta_i = \Theta_R$. The second is the familiar form of Snell's law

$$n_1 \sin \theta_i = n_2 \sin \theta_t \tag{30}$$

Total internal reflection occurs at the interface between two media where n_2 is less than n_1 and the angle of incidence is large. This happens in a step-index fiber at the core-cladding interface.

As Θ_i the angle of incidence is increased, Θ_t, the transmitted angle also increases. When $n_1 > n_2$, Θ_t is greater than Θ_i. There is a critical incidence angle for which $\Theta_t = 90°$. For incident angles greater than this critical angle no light propagates into medium 2.

The propagation vector in medium 2 is

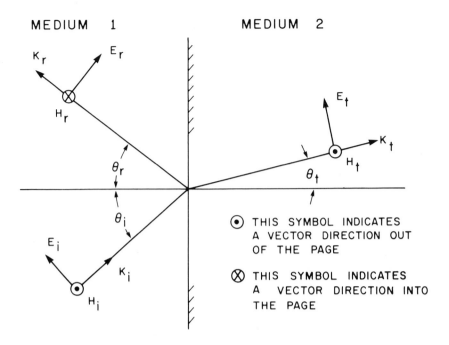

FIGURE 1. Plane wave reflection from a dielectric interface is represented in terms of the E, H, and K vectors of the incident reflected and transmitted waves. The plane of the paper is the plane of incidence (defined as the plane containing the surface normal and the incident ray). The polarization is parallel since the E field is parallel to the plane of incidence.

$$\vec{k}_t = n_2 k_o (\sin \theta_t \, \hat{x} + \cos \theta_t \, \hat{z}) \tag{31}$$

where by Snell's Law and a well-known trigonometric identity $\cos \Theta_t = [1 - (n_1/n_2)^2 \sin^2 \Theta_i]^{1/2}$. For incident angles greater than the critical angle, the $\cos \Theta_t$ and therefore the z component of k_t are imaginary. This results in an evanescent transmitted wave. The transmitted wave does not propagate in the z direction, but decays exponentially with z. The expression for the electric field associated with the wave in medium 2 is

$$\vec{\mathcal{E}}_t = \vec{E}_t \, e^{-z/\delta} \, e^{-j(n_2 k_o \sin \theta_t)x} \tag{32}$$

where $\vec{E}_t$ is a constant vector, and $\delta = (1/k_o)(n_2^2 - n_1^2 \sin^2 \Theta_1)^{-1/2}$ is the depth of penetration of the light into the second medium.

2. The Reflection and Transmission Coefficients

The reflected and transmitted waves are found in terms of the incident wave by applying boundary conditions to the interface. The tangential E and H fields are continuous across the interface. For parallel polarization shown in Figure 1, equating the tangential electric field in medium 1 to the tangential field in medium 2 results in the following equation,

$$E_i \cos \theta_i + E_R \cos \theta_R = E_t \cos \theta_t \tag{33}$$

Similarly, equating the tangential components of the magnetic fields results in the following equation

$$H_i - H_R = H_t \tag{34}$$

Equations 33 and 34 apply only to parallel polarization. A similar set of equations can be obtained for perpendicular polarization.

Since E and H for each of the three waves are related by the characteristic impedance of the medium, Equation 34 can be written in terms of the electric fields as follows,

$$n_1(E_i - E_R) = n_2 E_t \tag{35}$$

Multiplying Equation 33 by n_2 and using Equation 35 to eliminate E_t from the right hand side of Equation 33, results in the following

$$(E_i + E_R) n_2 \cos \theta_i = (E_i - E_R) n_1 \cos \theta_t \tag{36}$$

Solving for the reflection coefficient, defined as the ratio of E_R to E_i,

$$\rho_\| = \frac{E_R}{E_i} = \frac{n_1 \cos \theta_t - n_2 \cos \theta_i}{n_1 \cos \theta_t + n_2 \cos \theta_i} \tag{37}$$

Since by Snell's law and a well-known trigonometric identity, $\cos \theta_t = [1 - (n_1/n_2)^2 \sin^2 \theta_i]^{1/2}$ it follows that

$$\rho_\| = \frac{n_1[1 - (n_1/n_2)^2 \sin^2 \theta_i]^{1/2} - n_2 \cos \theta_i}{n_1[1 - (n_1/n_2)^2 \sin^2 \theta_i]^{1/2} + n_2 \cos \theta_i} \tag{38}$$

This equation is good for polarization parallel to the plane of incidence. A similiar analysis for polarization perpendicular to the plane of incidence yields the following expression for the reflection coefficient

$$\rho_\perp = \frac{\cos \theta_i - [(n_2/n_1)^2 - \sin^2 \theta_i]^{1/2}}{\cos \theta_i + [(n_2/n_1)^2 - \sin^2 \theta]^{1/2}} \tag{39}$$

Brewster's angle is the incident angle at which no parallel polarized light is reflected at a dielectric interface. When this occurs, the numerator of Equation 38 equals zero. That is,

$$n_1[1 - (n_1/n_2)^2 \sin^2 \theta_B]^{1/2} = n_2 \cos \theta_B \tag{40}$$

where θ_i has been replaced by θ_B, Brewster's angle. Squaring both sides of Equation 40, solving for the $\cos \theta_B$ using the trigonometric identity $\sin^2\theta = 1 - \cos^2\theta$ results in the following expression for the tangent of θ_B

$$\text{Tan } \theta_B = n_2/n_1 \tag{41}$$

The transmission coefficient is the ratio of the phasor representing the transmitted wave to the phasor representing the incident wave, E_t/E_i. It can be found using the boundary conditions in a manner similar to that used to obtain the reflection coefficients. The transmission coefficient for parallel polarization is

$$T_\| = \frac{2n_1 \cos \theta_i}{n_2 \cos \theta_i + n_1 [1 - (n_1/n_2)^2 \sin^2\theta_i]^{1/2}} \tag{42}$$

The transmission coefficient for perpendicular polarization is

$$T_\perp = \frac{2 \cos \theta_i}{\cos \theta_i + [(n_2/n_1)^2 - \sin^2 \theta_i]^{1/2}} \tag{43}$$

III. MATRIX RAY OPTICS

Ray propagation analysis provides a useful description in many situations. In isotropic materials, the ray direction is the direction of energy propagation. Rays are related to the wave description of optics in that rays are normal to constant phase surfaces. The matrix description of ray propagation allows complex optical structures to be described as combinations of simple elements. The matrix describing the structure is the product of the matrices of its elements. Below, the ray matrices for three optical elements that occur frequently in fiber optic systems are found.

A. Homogeneous Medium

A ray propagating through a homogeneous medium follows a straight line as shown in Figure 2. The displacement and slope of the line relative to the optic axis at reference plane 2 is described by the following set of linear equations

$$r_2 = r_1 + dr_1' \tag{44}$$

$$r_2' = r_1'$$

where r_2 and r_2' are the ray displacement and slope relative to the optic axis at reference plane 2. r_1 and r_1' are the ray position and slope at reference plane 1. The distance between reference planes is d. When matrices are used, Equation 44 becomes

$$\begin{bmatrix} r_2 \\ r_2' \end{bmatrix} = \begin{bmatrix} 1 & d \\ 0 & 1 \end{bmatrix} \begin{bmatrix} r_1 \\ r_1' \end{bmatrix} \tag{45}$$

B. Thin Lens

The thin lens is an idealized model that provides an accurate approximation to actual lenses. Paraxial rays are those rays propagating in directions nearly parallel to the axis. A thin lens changes the slope of a paraxial ray propagating through it an amount proportional to the displacement of the ray from the optical center of the lens. Actual lenses not obeying this law are said to have aberrations. Ray displacement from the axis is unchanged by the thin lens. Ray propagation from reference plane 1 just before a thin lens to reference plane 2 just after the thin lens is depicted in Figure 3 and is described by the following equation,

$$\begin{bmatrix} r_2 \\ r_2' \end{bmatrix} = \begin{bmatrix} 1 & 0 \\ -1/f & 1 \end{bmatrix} \begin{bmatrix} r_1 \\ r_1' \end{bmatrix} \tag{46}$$

where f is the focal length of the lens.

C. Quadratic Index Medium

A medium whose index of refraction varies as the square of the distance from the optical axis has the ability to guide rays. Such a medium is an idealized model for graded-index fibers. The index of refraction for such a medium is

$$n = n_o [1 - 1/2 \, (r/a)^2] \tag{47}$$

where n_o is the index of refraction on the optical axis, r is the distance from the optical axis, and a is a constant. Ray trajectories in this type of medium can be determined by applying

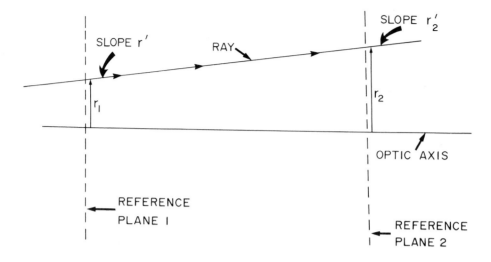

FIGURE 2..   Ray propagation through a homogeneous medium is shown. A ray follows a straight line in such a medium.

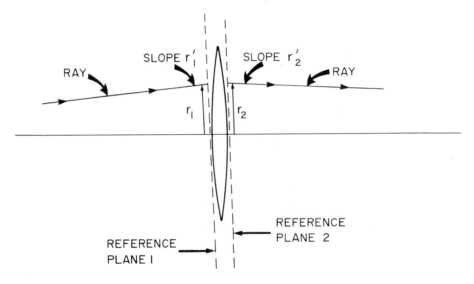

FIGURE 3.   Ray propagation through a thin lens is shown. The displacement of the ray from the axis is unchanged by the lens. The ray slope is changed by an amount proportional to the ray displacement from the axis.

Snell's law to the ray as it propagates through the medium as shown in Figure 4. Snell's law applied to this case is

$$n(r) \cos\theta = \left[n(r) + \frac{\partial n}{\partial r} dr \right] \cos(\theta + d\theta) \tag{48}$$

Equation 48 may be simplified by using the trigonometric identity for the cosine of the sum of two angles and assuming, $\cos(d\Theta) = 1$ and $\sin(d\Theta) = d\Theta$. This results in the following equation:

$$\frac{dn}{dr} = n \tan\theta \frac{d\theta}{dr} \tag{49}$$

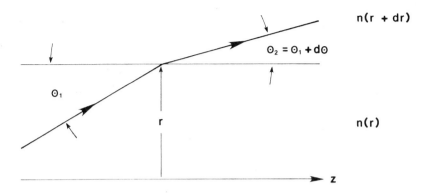

FIGURE 4. A ray propagating in an inhomogenous medium is shown. The index of refraction is a function of r, the distance from the z axis. When the ray moves to a region with a different index of refraction its slope is changed.

The tangent of Θ is the slope of the ray; that is, $\tan \Theta = dr/dz$. Therefore,

$$\frac{dn}{dr} = n \frac{dr}{dz} \frac{d\theta}{dr} \tag{50}$$

$$= n \frac{d\theta}{dz}$$

For paraxial rays, Θ is the slope of the ray; that is, $\Theta \simeq dr/dz$. Therefore,

$$\frac{dn}{dr} = n_o \frac{d^2r}{dz^2} \tag{51}$$

where n is assumed to be approximately equal to n_o in the right-hand side of Equation 51. When Equation 47 is used in Equation 51, the following equation results,

$$\frac{d^2r}{dz^2} = -\frac{r}{a^2} \tag{52}$$

where r is the displacement from the optic axis. The solution to Equation 52 is a sinusoid and represents an oscillation of the ray about the optical axis as it propagates. The ray displacement and slope relative to the optic axis as a function of distance along the optic axis are given by the following matrix expression

$$\begin{bmatrix} r(z) \\ r'(z) \end{bmatrix} \begin{bmatrix} \cos(z/a) & a\sin(z/a) \\ -\frac{1}{a}\sin(z/a) & \cos(z/a) \end{bmatrix} \begin{bmatrix} r(o) \\ r'(o) \end{bmatrix} \tag{53}$$

where r(o) and r'(o), are the ray displacement and slope at the input reference plane.

The quadratic index medium is not only an idealized description for graded-index fibers but also for the GRIN rods discussed in Chapter 7.

IV. GAUSSIAN BEAMS

Optical beams produced by laser resonators and propagation modes in quadratic index media can be described using Gaussian beams. These beams represent an approximate solution to the wave equation. They are, nevertheless, an accurate representation of laser beams. These beams differ from uniform plane waves in that the energy is confined to a narrow region about the axis, and the constant phase surfaces are spherical rather than plane. A component of a field quantity associated with the beam satisfies the scalar wave equation.

$$\nabla^2 E + k^2 E = 0 \qquad (54)$$

where k is the propagation constant of the medium. For a beam traveling in the z direction,

$$E = \psi(x,y,z) \, e^{-jkz} \qquad (55)$$

where ψ is a slowly varying complex function representing the difference between a laser beam and a plane wave. That is, ψ describes the non-uniform intensity distribution, nonplanar wavefronts, and the expansion of the beam with distance of propagation.

Substituting Equation 55 into Equation 54 results in the following equation,

$$\frac{\partial^2 \psi}{\partial x^2} + \frac{\partial^2 \psi}{\partial y^2} + 2jk \frac{\partial \psi}{\partial z} = 0 \qquad (56)$$

where it has been assumed that $\partial^2 \Psi / \partial z^2$ can be neglected because Ψ varies slowly with z. A solution of Equation 56 is

$$\psi = e^{-j\left(P + \frac{kr^2}{2q} \right)} \qquad (57)$$

where $r^2 = x^2 + y^2$ and, P and q are functions of z, P(z) is a complex phase shift, and q(z) is the beam q parameter. As will be seen below, this q parameter can be used to calculate the beam spot size and phase front curvature. Beam spot size and phase fronts change as the beam propagates. These beam changes can be calculated by determining the variation of q with z. Substituting Equation 57 into Equation 56 and comparing terms of equal powers of r results in the following relation.

$$\frac{dq}{dz} = 1 \qquad (58)$$

It follows from Equation 58 that

$$q_2 = q_1 + z \qquad (59)$$

where q_2 is the q parameter at one plane (output plane), and q_1 is the q parameter at a second plane (input plane). The displacement of the second plane from the first is z. Equation 59 expresses the transformation of the q parameter, and, therefore, the beam spot size as the beam propagates through a uniform medium.

Spot size control is important in fiber optic situations such as coupling optical energy into and out of fibers. Spot size and phase front curvature are related to the q parameter by the following equation:

$$\frac{1}{q} = \frac{1}{R} - j \frac{\lambda}{\pi \omega^2} \qquad (60)$$

where R is the radius of curvature of the phase fronts, and ω is the spot size. Substituting Equations 60 and 57 into Equation 55 and recalling $k = 2\pi/\lambda$

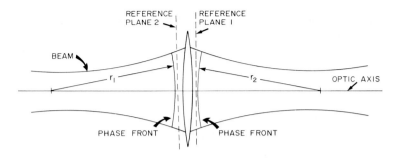

FIGURE 5. The transformation of a Gaussian beam by a thin lens is shown. The lens changes the radius of curvature of the beam phase fronts from r_1 to r_2 but not the beam spot size.

$$E(x,y,z) = E(z)e^{-\left(\frac{r}{\omega}\right)^2} e^{-j\frac{kr^2}{2R}} \tag{61}$$

The magnitude of field quantities is a Gaussian function of r, the radial coordinate. At a distance from the axis of ω, field quantities have a value equal to e^{-1} of their own axis value. This defines the spot size ω.

This solution is only one of a set of functions that satisfy Equation 56. It is the solution associated with the lowest order transverse mode. Higher order transverse modes contain Hermite polynomial functions of r as factors.[14] Deviations from the simple Gaussian transverse energy distribution can be attributed to the presence of higher order modes.

A. Thin Lens Beam Transformation

To describe beam propagation through optical structures, it is necessary to determine the transformation of a Gaussian beam by a thin lens.

Since rays are perpendicular to constant phase surfaces and the radius of curvature is also perpendicular to the spherical constant phase surfaces, a ray can be considered to propagate along the radius of curvature of the spherical phase surfaces. Consider Figure 5. To the left of the thin lens, the ray directed along the radius of curvature has a slope x/R_1, where x is the displacement of the ray from the axis at the lens and R_1 is the radius of curvature of the phase front. After passing through the lens, the slope is x/R_2, where R_2 is negative. Phase fronts with enters of curvature to the right of the front are considered to have a negative radius of curvature.

The change in slope is found by applying Equation 46 to this situation:

$$\frac{x}{R_2} - \frac{x}{R_1} = -\frac{x}{f} \tag{62}$$

Since $1/q = 1/R - j\,\lambda/\pi\omega^2$ and since this spot size, ω, is unchanged by a thin lens, the tranformation of the q parameter by a thin lens is

$$\frac{1}{q_2} = \frac{1}{q_1} - \frac{1}{f} \tag{63}$$

Note that for both propagation through a thin lens and for propagation through a uniform medium

$$q_2 = \frac{Aq_1 + B}{Cq_1 + D} \tag{64}$$

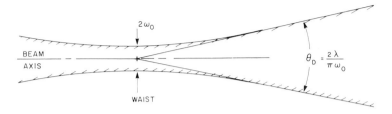

FIGURE 6.   A laser beam is shown. The angle of divergence describes the increase in beam spot size (radius), ω, with distance of propagation far from the waist.

where A, B, C, and D are the terms of the ABCD matrix describing ray propagation. Equation 64 is general and applies not only to propagation through uniform media and thin lenses but to other optical structures as well.

B. Beam Numerical Aperture

The angle of divergence of a beam describes the increase in spot size with distance of propagation. This situation is shown in Figure 6. The angle of divergence is $2 \, \mathrm{Tan}^{-1}(\omega/z)$ where z, the distance from the beam waist, is large. The angle of divergence, Θ_D, depends on the beam spot size at the waist. At the waist, the phase fronts are plane, that is, spherical with infinite radius of curvature, and the q parameter is imaginary.

$$q_o = j \frac{\pi \omega_o^2}{\lambda} \tag{65}$$

where ω_o is the spot size at the waist. After propagating through a distance z, $q_2 = q_o + z$, and

$$\frac{1}{q_2} = \frac{1}{z} - j \frac{\pi \omega_o^2}{\lambda z^2} \tag{66}$$

This is true for $z \gg \omega_o^2/\lambda$, since from the definition of the q parameter (Equation 60)

$$\frac{1}{q_2} = \frac{1}{R} - \frac{j\lambda}{\pi \omega^2} \tag{67}$$

a comparison of Equations 66 and 67 indicates that at a distance far from the waist, the radius of curvature of the phase fronts is equal to the distance from waist and also the spot size,

$$\omega = \frac{\lambda z}{\pi \omega_o}$$

The angle of divergence of the beam is $2 \, \mathrm{Tan}^{-1} (\omega/z)$

$$\theta_D = \frac{2\lambda}{\pi \omega_o} \tag{68}$$

where it has been assumed Θ_D is small and the tan Θ_D equals Θ_D. Sometimes the term numerical aperture (NA) is used to describe the divergence of a beam. The NA is defined as the sine of Θ_D. For small Θ_D the NA is given by

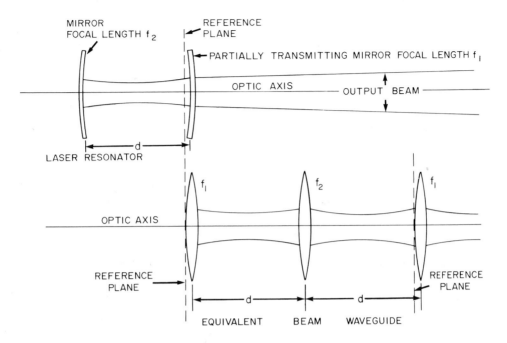

FIGURE 7. A laser resonator consisting of two mirrors separated by a distance, d, and its beam waveguide equivalent are shown.

$$NA = \frac{\lambda}{\pi\omega_o} \tag{69}$$

C. Laser Resonator Beams

The spot size and angle of divergence of a beam produced by a laser is determined by the resonator ABCD matrix. Consider a laser resonator consisting of two spherical mirrors separated by a distance, d, as shown in Figure 5. Optical energy is stored in the resonator in the form of electromagnetic waves reflected back and forth between the mirrors. One mirror is partially transmitting. The beam emitted has an angle of divergence and spot size that can be calculated from the ABCD matrix description of the resonator optics. In actual laser resonators, the medium between the mirrors providing the optical gain is not homogeneous. This is particularly true for semiconductor lasers where the medium provides guiding and confinement of the beam.

A ray propagating in the beam waveguide equivalent to a laser resonator is shown in Figure 7. The resonator, shown in Figure 7, is assumed to have a homogeneous medium between the mirrors. The beam waveguide equivalent has lenses with focal lengths equal to the focal lengths in the laser resonator. The distance between lenses is equal to the resonator mirror separation. A ray propagating in the resonator is focused and reflected when it strikes a mirror. The equivalent ray in the beam waveguide is focused and transmitted when it passes through a lens.

The ABCD matrix description of ray propagation through one iteration of the beam waveguide is

$$\begin{bmatrix} r_{n+1} \\ r'_{n+1} \end{bmatrix} = \begin{bmatrix} A & B \\ C & D \end{bmatrix} \begin{bmatrix} r_n \\ r'_n \end{bmatrix}$$

$$= \begin{bmatrix} (1 - 2d/f_1 - d/f_2 + d^2/f_1f_2) & d(2 - d/f_2) \\ (d/f_1f_2 - 1/f_1 - 1/f_2) & (1 - d/f_2) \end{bmatrix} \begin{bmatrix} r_n \\ r'_n \end{bmatrix} \tag{70}$$

where r_n and r'_n are the ray position and slope at the nth reference plane. From the first row of the matrix equation it follows that

$$r'_n = \frac{1}{B} r_{n+1} - \frac{A}{B} r_n \tag{71}$$

The use of Equation 71 allows ray slopes to be expressed in terms of ray displacement. It follows from the second row of Equation 70 that $r'_{n+1} = Cr_n + Dr'_n$. When Equation 71 is used to eliminate slope terms from the second row of Equation 70, the following equation results:

$$r_{n+2} - (A + D)r_{n+1} + r_n = 0 \tag{72}$$

Equation 72 is the difference equation describing the beam trajectory in the beam waveguide and, therefore, in the laser resonator. The coefficient of r_n is the determinant of the ABCD matrix. This coefficient is one that is, $AD - BC = 1$. Justification for this can be seen by noticing that the determinants of the ABCD matrices describing propagation through space and through a thin lens are unity. Also, the determinant of the product of two ABCD matrices is the product of the determinants of the individual matrices. Therefore, the determinant of the ABCD matrix describing any arbitrary sequence of lenses and spaces will have a determinant equal to unity.

Solutions of Equation 72 that result in a r_n that goes to infinity with increasing n represent unstable resonators. The conditions for resonator stability can be found by substituting $r_n = \lambda^n$ in Equation 72. Here lambda is a constant that depends on the resonator geometry. It is not the wavelength. For this case it follows from Equation 72

$$\lambda^2 + (A + D)\lambda + 1 = 0 \tag{73}$$

Lambda is less than one and, therefore, the resonator is stable if

$$\left| \frac{A + D}{2} \right| \leq 1 \tag{74}$$

Beam spot size at the resonator output mirror and the angle of divergence of the emitted beam can be calculated from the ABCD matrix describing ray propagation for one round trip pass through the laser resonator. Assume a selfconsistent beam. That is, assume the beam q parameter, and, therefore, the beam geometry is unchanged after propagating through one round trip in the resonator and returning to its original reference plane. Using the ABCD law, Equation 64, and assuming the q parameter to be unchanged by one round trip in the resonator results in the following

$$q = \frac{Aq + B}{Cq + D} \tag{75}$$

where q is the q parameter at the output reference plane.

Equation 75 can be solved to obtain the following expression for 1/q

$$\frac{1}{q} = \frac{A - D}{2B} - j \frac{\left[1 - \left(\frac{A + D}{2} \right)^2 \right]^{1/2}}{B} \tag{76}$$

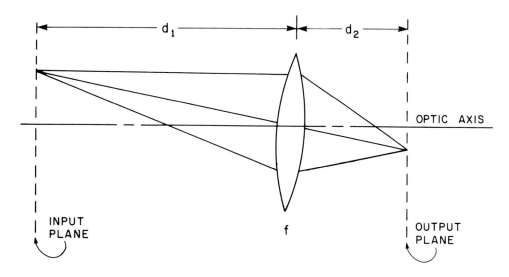

FIGURE 8. Single lens imaging system.

The output beam spot size can be calculated using Equation 60 and Equation 76. The angle of divergence of the output beam is obtained by using Equation 59 to determine how the q parameter changes with a distance far from the output plane.

PROBLEMS

1. A plane wave propagating in a glass with an index of refraction of 1.5 is reflected when it is incident on an interface with a second glass whose index of refraction is 1.4. The wavelength is 0.82 μm. What is the critical angle for total internal reflection? If the angle of incidence is twice the critical angle, what is the depth of penetration into the second medium?

2. Calculate Brewster's angle for light propagating in air and incident on glass with an index of refraction of 1.5.

3. An optical detector is coated with a glass of index of refraction 1.5. Calculate the power loss due to reflection at the air glass interface for a plane wave normally incident on the detector.

4. Determine the ABCD matrix, relating ray position and slope at the image plane to ray position and slope at the object plane for the single lens imaging system shown in Figure 8. Imaging occurs when all rays leaving a single point in the object plane, regardless of their slope, arrive at a single point in the image point as shown in Figure 8. For this to happen the B term in the ABCD matrix must be zero. Show that if the B term is zero $1/d_1 + 1/d_2 = 1/f$.

5. Find the ABCD matrix for ray propagation through a dielectric interface. Assume paraxial rays, i.e., $\sin \Theta = \Theta$, where Θ is the angle of incidence.

6. Find the spot size at the output of plane of the laser resonator shown in Figure 7. The focal lengths of mirrors one and two are 1 and 2 m, respectively. The mirrors are separated by 50 cm. What is the angle of divergence of the output beam?

REFERENCES

1. **Kao, K. C. and Davies, T. W.,** Spectrophotometric studies of ultra low loss optical glasses 1: single beam method, *J. Sci. Intrum. (J. Phys. E),* Series 2, 1, 1063, 1968.
2. **Kapron, F. P., Keck, D. B., and Maurer, R. D.,** Radiation losses in glass optical waveguides, *Appl. Phy. Latt.,* 17 (10), 423, 1970.
3. **Nakachar, M. and Inagaki, N.,** Fabrication of a 100 km graded-index fiber from a continuously consolidated VAD preform, Digest of Technical Papers, Third Int. Eng. Conf. on Integrated Optics and Optical Fiber Communication OSA/IEEE, San Francisco, April 27 to 29, 1981, 100.
4. **Furui, Y., Kamiza, T., Ueki, A., and Sentsui, S.,** Fabrication study of single-mode optical fibers with P/F cladding, Digest of Technical Papers, Third Int. Conf. on Integrated Optics and Optical Fiber Communication, OSA/IEEE, San Francisco, April 27 to 29, 1981, 100.
5. **Basch, E. E., Beaudette, R. A., and Carnes, H. A.,** Optical transmission for interoffice trunks, *IEEE Trans. Commun.,* COM-26 (7), 1007, 1978.
6. **Giallorenzi, T. G.,** Optical Communications research and technology: fiber optics, *Proc. IEEE,* 66(7), 744, 1978.
7. **Cook, J. S.,** Bell system lightwave links, Digest of Technical Papers, Third Int. Conf. on Integrated Optics and Optical Communication, OSA/IEEE, San Francisco, April 27 to 29, 1981, 10.
8. **Runge, P. K.,** Future transatlantic fiber optical communication systems, Digest of Technical Papers, Third Int. Conf. on Integrated Optics and Optical Communications, OSA/IEEE, San Francisco, April 27 to 29, 1981, 20.
9. **Deichmiller, A. C.,** Progress in fiber optics transmission systems for cable television, *IEEE Trans Cable Television,* CATV-5(2), 50, 1980.
10. **Lacy, E. H.,** *Fiber Optics,* Prentice-Hall, Englewood Cliffs, N.J., 1982, 11.
11. **Crow, J. D., and Sachs, M. W.,** Optical fibers for computer systems, *Proc. IEEE,* 68(10), 1275, 1980.
12. **Barnoski, M. K.,** Fiber systems for the military environment, *Proc. IEEE,* 68(10), 1315, 1980.
13. **Allard, F. C., Olin, L. D., and Manes, E. F.,** A Fiber Optic Sonar Link: Fiber Optic Components Design Considerations and Development Status, Proc. EO/Laser '80 Conf., Boston, November 19 to 21, 1980, 114.
14. **Kogelnik, H. and Li, T.,** Laser beams and resonators, *Proc. IEEE,* 54(10), 1312, 1966.

Chapter 2

OPTICAL WAVEGUIDE MANUFACTURING

Dr. Ishwar D. Aggarwal

TABLE OF CONTENTS

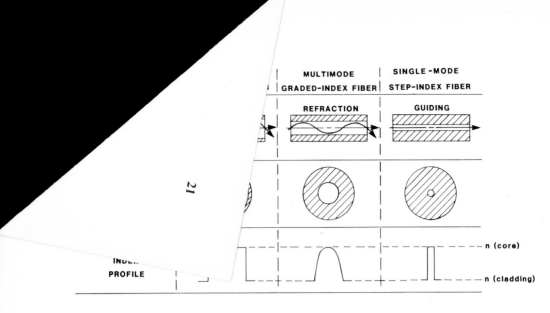

FIGURE 1. The fundamental types of optical fibers.

I. INTRODUCTION

The transmitting medium, to a large extent, is the most important component in the optical fiber communication system. It is the rapid development of a low-loss, high-bandwidth optical fiber or waveguide that has made optical fiber communication a reality today. Recently, fiber with optical loss as low as 0.16 dB/km at 1500 nm has been reported.[1] Using glass fibers, optical telecommunication systems as long as 100 km, without the use of repeaters, have been demonstrated.[2]

An optical fiber is fabricated from two compositions of glasses. One has a relatively high refractive index and forms the core of the fiber. The core is surrounded by the second composition that has a slightly lower refractive index and forms the cladding portion of the fiber. As shown in Figure 1, light is guided in the core by means of total internal reflection, internal refraction, or by internal guiding.

There are three basic types of optical fibers: multimode step-index, multimode graded-index, and single-mode step-index. These fiber types are displayed in Figure 1. Multimode step-index fiber is typically aimed for low data-rate transmission. Multimode graded-index, on the other hand, can be used for high data-rate transmission, due to the minimization of intermodal pulse broadening. Intermodal pulse broadening results from the differences in path travel times for the various modes of light guided by the fiber. This decrease in pulse broadening is achieved through proper grading of the refractive index profile of the optical fiber.[3] The core diameters for multimode graded- and step-index fibers are approximately 50 to 100 μm. The peak difference in refractive indices of the core and the cladding is between 1 to 2%. This index difference determines the numerical aperture of the fiber, which in turn determines the amount of coupled light the fiber can accept from an optical source.

The single-mode type of fiber, as shown in Figure 1, has a core diameter of about 5 to 10 μm and supports only one mode of light. This fiber type offers the ultimate in high bandwidth because there is no intermodal pulse broadening in a single-mode waveguide. The total dispersion in single-mode fibers is a result of material dispersion and waveguide dispersion. In certain cases, these two types of dispersion can be of opposite sign and of similar magnitude, thus effectively canceling each other out over a narrow wavelength region, resulting in zero-dispersion waveguides.[4] Fibers have been fabricated with zero dispersion at various wavelengths in the 1.3 to 1.5 μm region.[5]

There are many manufacturing processes for making single-mode and multimode fibers. The manufacturing process basically consists of four main operations.

1. Preform fabrication
2. Fiber drawing
3. Optical characterization
4. Mechanical characterization

In preform fabrication, either one glass rod consisting of both the core and cladding is prepared, or glass materials for the core and the cladding are separately prepared. The preform glass is then drawn into an optical fiber of the desired dimensions. The optical poperties of the drawn fiber are then measured. Optical fibers are screened to insure minimum acceptable strengths. The minimum acceptable strength has been determined from mechanical properties such as tensile strength and static fatigue. This data is used to forecast expected survival times of optical fibers.

II. PREFORM FABRICATION

There are many manufacturing processes for making single and multimode fibers. The selection of the fabrication process for making optical fiber is dependent upon the desired type of fiber and its properties. The important properties of an optical waveguide are as follows:

1. Spectral optical attenuation
2. Numerical aperture
3. Spectral bandwidth or pulse dispersion
4. Tensile strength and static fatigue

The following properties are important for single-mode fibers:

1. Spectral optical attenuation
2. Zero dispersion wavelength
3. Spot size
4. Cut-off wavelength
5. Tensile strength and static fatigue

Optical attenuation occurs due to three distinct mechanisms: material absorption, scattering, and waveguide losses. Material absorption in the usable wavelength region is the result of transition metal impurities such as Fe, Co, Ni, Cu, Cr, Mn, and V in the glass composition. As shown in Figure 2, considerable signal loss can occur with only a few parts per billion of these impurities present.[6] Scattering loss takes place due to inherent equilibrium density and composition fluctuations (Rayleigh scattering). In addition, considerable scattering can occur due to large inhomogenieties, bubbles, and crystallites in the fiber. This type of loss, however, can be eliminated by good process control. Waveguide losses can also occur due to irregularities and variation at the core/clad interface of the fiber. Again, this type of loss can be reduced significantly with process control. Absorption loss is the major loss and can only be reduced by careful material and process selection.

The glass materials selected for making communication fibers should be available in pure form and with appropriate refractive indices. It is necessary to be able to change and control the refractive index of these materials in order to achieve a suitable index profile for minimum pulse dispersion or high bandwidth in the case of graded-index multimode fibers.

The materials should also have low dispersion and Rayleigh scattering. The core and the cladding glasses should have appropriate thermal expansion coefficients and viscosities in

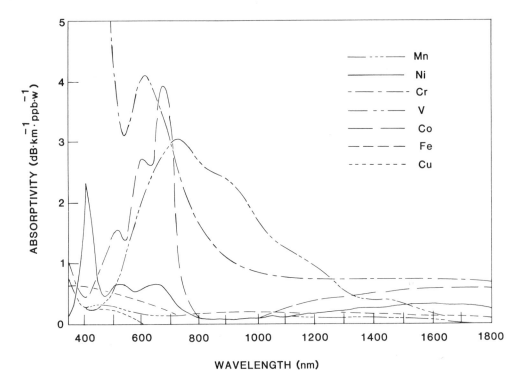

FIGURE 2.   Absorptivity of transition metal ions vs. wavelength.

order to make both preforms and fiber with the proper amount of residual stresses. Finally, the materials should have high chemical durability and mechanical strength. There are few materials that meet all of these requirements. A compromise between fabrication processes and materials is required in order to make fibers for various applications.

Described below are some of the processes available for manufacturing preforms for optical fibers.

1. *Rod and Tube Process.* This is probably the oldest and simplest process for making fibers. In this process, a cleaned and polished rod of the desired core composition is inserted into a cleaned tube of the cladding glass composition. The composite is then drawn into a fiber of appropriate diameter using a furnace.[7] Gas bubbles and dirt may be entrapped in the core-cladding interface. This will give rise to additional losses. The main advantage in this process is that fibers with a high numerical aperture (NA) can be made at very low costs. Fibers with losses of 400 dB/km at 850 nm wavelength and NA of 0.6 are being made using this process. This process can only produce step-index multimode fibers and is not commonly used for making low-loss fibers.

2. *Double Crucible Process.* In this process, the core and the cladding glasses are contained in two separate concentric crucibles with an orifice in the bottom of each.[8] As displayed in Figure 3, the two crucibles are heated in a furnace until the glasses are soft enough to flow through the orifice and be drawn into fiber. This manufacturing process can be run continuously. Tremendous progress with this process has been made recently by the British Post Office in the U.K.[9] Step- and graded-index fibers with losses of less than 5 dB/km at 850 nm wavelength have been fabricated. The main disadvantages in this process are that it requires a clean room facility to avoid impurities, and only low silica content glasses, which have lower chemical durability, are used.

3. *Float Process.* This process was developed by Pilkington Brothers in the U.K. Cladding

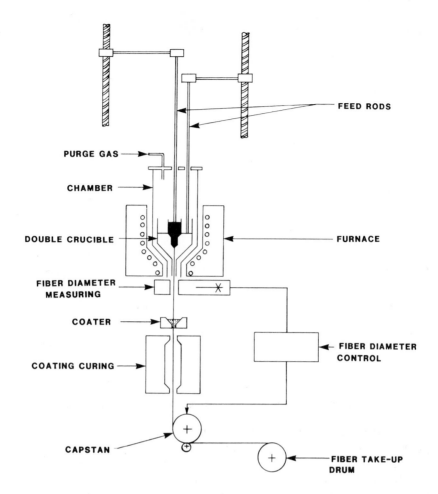

FIGURE 3. Double crucible fiber manufacturing process.

glass is floated on the top of core glass in a fused silica crucible. A rod is drawn from the core glass in such a way that it is coated by cladding glass as it passes out of the crucible. These cladded rods are then drawn into fibers of appropriate diameters. Fibers containing low silica, which are not as chemically durable, are made by this process. Step-index fibers with losses of about 50 dB/km and NA of about 0.45 are produced in the float process.

4. *Laser Process to Draw Plastic Clad Silica Fibers.* Fused silica is a highly durable single component glass having a very low Rayleigh scattering loss. One of the less expensive methods for making optical fiber is to draw a fused silica core and clad it with a low refractive index plastic. Such fiber is called a plastic cladded silica (PCS) fiber.[10]

In the earlier process, fibers were made using low silica content glasses. These glasses have a low softening temperature, and can be drawn into fiber using ordinary furnaces. For drawing fused silica or high silica content glass fibers, a temperature in excess of 2000°C is required.

To draw such fibers, a high temperature source such as a hydrogen-oxygen burner, an r.f. induction furnace, a tungsten or carbon resistance furnace, or a CO_2 laser must be used. In order to keep optical losses to a minimum in the plastic cladded silica fibers, it is also important to use a clean heat source. Generally, a CO_2 laser (250 W) or modified resistance furnace is used to draw fused silica fibers. The drawn silica fiber is coated with a silicone type of material that acts as the cladding. The cladded fiber may then be jacketed with a thermo-plastic such as Nylon® or Hytrel®. Fibers with losses of about 10 dB/km at 820 nm wavelength, and NA of 0.35 are made by this process.

5.   *Phasil Process*. This process was invented and developed by T. A. Litovitz and P. B. Macedo at Catholic University.[11] Rods of a sodium-borosilicate glass of suitable composition are prepared, using reagent grade materials. This glass is then separated at an elevated temperature into low and high silica content phases. The low silica phase contains most of the metal impurities. This phase is then leached out using an acidic solution. The remaining leached rod consists of a porous glass material with a composition containing about 96% of SiO_2. This rod is stuffed with a $CsNO_3$ solution. The lution is then diffused out of the rod in water to give the appropriate concentration profile. This structure is sintered to yield a glass preform with a corresponding refractive index profile. The preform is drawn into fiber using a high temperature heat source. Step- and graded-index fibers with about 10 dB/km loss at 850 nm wavelength and an NA of 0.3 have been prepared using this process.

6.   *Flame Hydrolysis (OVPO)*. In most of the processes previously described, expensive, high-purity, solid raw materials are used to make step- and graded-index fibers. Moreover, the graded-index fibers made by these processes do not meet high bandwidth (—1 GHz) requirements.

   Flame hydrolysis,[12-13] also known as outside vapor phase oxidation (OVPO), or the soot process (developed by Corning Glass Works), produces very low-loss (≤2 dB/km at 820 nm wavelength) fibers with high bandwidths and reasonable numerical apertures. As shown in Figure 4, in this technique, vapors of purified chemical precursors, such as silicon tetrachloride, germanium tetrachloride, boron trichloride, and phosphorous oxychloride are passed through a burner and are hydrolyzed at high temperature to produce mixtures of glass-forming oxides of very high purity, as shown in the following reactions:

$$SiCl_4 + O_2 \rightleftarrows SiO_2 + 2Cl$$

$$GeCl_4 + O_2 \rightleftarrows GeO_2 + 2Cl_2$$

$$4BCl_3 + 3O_2 \rightleftarrows 2B_2O_3 + 6Cl_2$$

$$4POCl_3 + 3O_2 \rightleftarrows 2P_2O_5 + 3Cl_2$$

In the flame hydrolysis process, these reactions take place in an open environment, and the oxide mixture is deposited as finely divided soot on a rod. The rod is removed and the soot preform is subsequently consolidated to yield a clear glass preform from which the fiber is drawn. The main advantage of this process is that low-loss, step-, or graded-index fibers can be made with deposition rates of about 10 g/min. The main disadvantage is that the water that is formed as a reaction product in the flame contaminates the glass. This water contamination increases the optical attenuation of the fiber at wavelengths of 0.945 and 1.385 μm. However, this disadvantage has recently been overcome and fibers with losses as low as 0.16 dB/km at 1.55 μm have been reported.

7.   *Modified Chemical Vapor Deposition (MCVD)*. The overall modified chemical vapor deposition method is displayed in Figure 5. Most recently, scientists at Bell Laboratories[14] and elsewhere have used purified halides of silicon, germanium, phosphorus, and boron as raw materials in the preform fabrication process. Vapors of these materials in appropriate proportions are oxidized at high temperature and deposited as glass on the inside surface of a cleaned fused silica tube. After deposition of the cladding and

SOOT DEPOSITION

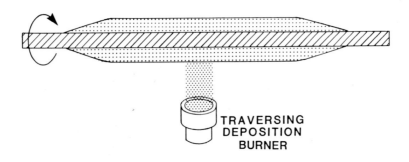

TRAVERSING
DEPOSITION
BURNER

O_2 & METAL HALIDE VAPOR

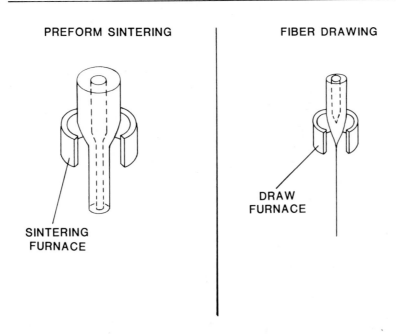

FIGURE 4.   Flame hydrolysis fabrication method or outside vapor phase oxidation (OVPO).

the core glass compositions, the tube is collapsed to form a rod. Fiber of appropriate size is then drawn from this rod using a high temperature ($\approx 2000°C$) heat source. The advantages of this process are

(1) The raw materials used are liquids and are available in high purity form.
(2) The raw materials are vaporized before being rejected. This provides an extra distillation or purification step.
(3) The vapor formation, chemical reaction and deposition take place in a closed system. This effectively excludes external contamination.

Oxygen is used as the carrier gas for the various chloride materials used. Figure 6 shows the attenuation vs. wavelength that can be achieved using the process to fabricate

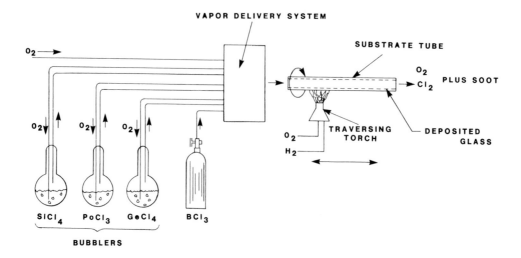

FIGURE 5.   Modified chemical vapor deposition fabrication process.

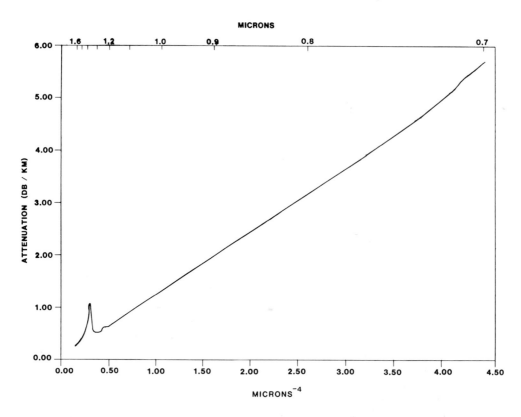

FIGURE 6.   Attenuation vs. wavelength for a graded index fiber fabricated using the MCVD process.

graded-index multimode fiber. Both step- and graded-index multimode, and single-mode fiber with very low loss (≤0.2 dB/km at 1300 nm wavelength) and several gigahertz bandwidth can be made using this process.

8.   *Vapor Axial Deposition (VAD).* In this process, developed by researchers at NTT Laboratories,[15] core glass and cladding glass soot are simultaneously deposited in an axial direction on a rotating mandrel to form a rod-like soot preform as shown in

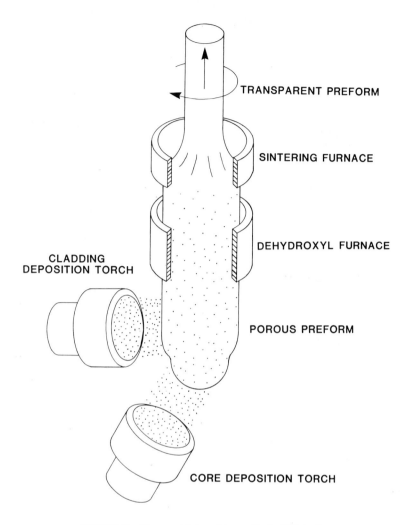

TRANSPARENT PREFORM

SINTERING FURNACE

DEHYDROXYL FURNACE

CLADDING
DEPOSITION TORCH

POROUS PREFORM

CORE DEPOSITION TORCH

FIGURE 7. Vapor axial deposition method of fabrication.

Figure 7. The soot preform is consolidated into clear glass either simultaneously as deposited or in a separate process step. The preform thus made is drawn into fiber. This process has the potential for making large preforms continuously, resulting in lower fiber costs. The fiber made with this process does not have a central index depression or hole, as is the case for the previous two processes. This process, although requiring rigorous process control, is most suited for single-mode or step-index fibers. Single and multimode fibers with very low losses (0.2 dB/km @ 1550 nm) have been made using this process, and single mode fibers as long as 100 km have been drawn from VAD preforms.

9. *Plasma Chemical Vapor Deposition (PCVD)*. This process has been introduced by and investigated in Philips Research Laboratories[16-18] during the past few years. In this process, volatile glass-forming reactants such as $SiCl_4$, $GeCl_4$, SiF_4, and BCl_3, etc. are introduced into a silica tube just as in the MCVD process. As shown in Figure 8, the essential difference is that in the PCVD process the oxide formation is stimulated by means of a nonisothermal plasma operating at 2.4 GHz frequency and a low working pressure on the order of 10 torr. In PCVD, complete oxide deposition occurs heterogeneously on the inner wall of the tube under low pressure. Thin layer deposition is achieved by traversing the plasma zone rapidly along the axis of the substrate tube.

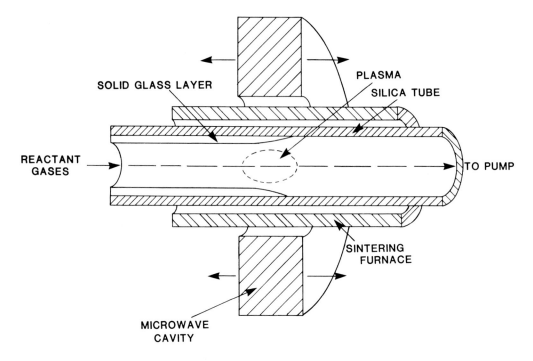

FIGURE 8.   Plasma chemical vapor deposition fabrication process.

Due to the high deposition efficiency, the composition of the layers can be adjusted accurately by carefully controlling the composition of the gas phase raw materials. Fibers with very low loss, 0.6 dB/km @ 1300 nm and high bandwidth (>1 GHz-km) have been routinely obtained from preforms made using this process.

Another advantage of this process is that relatively high concentration of fluorine can be incorporated into the deposited glass. This will allow for a central core glass of pure SiO_2 surrounded by a lower refractive index fluorine doped glass. Such a composition would result in even lower attenuation.

10.   *Plasma Enhanced (PEMCVD).* This is a variation of the MCVD process. In this technique, oxygen RF plasma is used to enhance the reaction of the chloride vapors to oxide soot at atmospheric pressure.[19-21] As shown in Figure 9, the soot is separately deposited and sintered as fine layers of glass on the inner wall of the substrate tube. The RF coil and burner traverse simultaneously, but independently, at optimized rates of speed. Using this process, deposition rates of 5 to 10 g/min have been achieved. This process, like the other vapor deposition processes, is capable of making low loss fibers and is most suited for step-index multimode or single-mode fibers.

Although all previously discussed fabrication methods are capable of producing optical fiber, the processes that involve vapor phase reactions (OVPO, MCVD, VAD, PCVD, etc.) are receiving the most attention in the commercial market. Generally in the vapor phase reaction, germanium and phosphorus doped silica comprise the core glass in the optical fiber. Boron and fluorine doped silica, which depresses the refractive index of silica, has been used quite successfully in the fibers' optical cladding. The use of boron in the optical core has resulted in increased absorption at wavelengths greater than 1300 nm. This region is of particular interest since the silica-based optical fibers typically exhibit the lowest attenuation at wavelengths between 1300 to 1550 nm.

Several key factors such as deposition rate, deposition efficiency, and fiber type currently govern the selection of a particular process for optical fiber fabrication. Currently, the

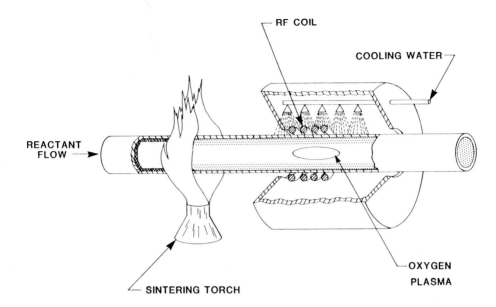

FIGURE 9. Plasma enhanced modified chemical vapor deposition process.

PEMCVD and OVPO processes exhibit the highest deposition rates of all fiber fabrication methods. High deposition efficiencies can be achieved, currently, by using the PCVD or PEMCVD process. The VAD process is a semicontinuous process and will have relatively low labor costs.

The ultimate fabrication process will depend upon the type of fiber to be fabricated and the interaction of deposition rate, deposition efficiency, and the resulting optical yield from the particular fabrication process. To date, there is no clear-cut choice for an optimum optical fiber fabrication process. Each process has its own advantages and disadvantages, with each being capable of making high quality high performance optical fibers.

The ability to implement process rate improvement, while maintaining a high yield in production, will establish the eventual predominance of one or more of these fabrication techniques.

III. FIBER DRAWING

This section describes how optical waveguide fibers are produced from the preforms described in the preceding section. First the basic concepts and principles of the process are described along with the inherent limitations and unique problems. These are then followed by detailed descriptions of the methods, procedures, and equipment.

The process by which optical waveguide preforms are transformed into fibers is called fiber drawing. In the drawing process, the preform is gripped at one end and suspended vertically. The lower end of the preform is heated until the glass softens. The soft viscous glass at the tip of the preform is then pulled and stretched into a fiber, maintaining the exact cross-section of the preform at a severely reduced scale. The process is made continuous by slowly and steadily moving the preform into the heating zone while the fiber continues to be pulled via a motorized pinch wheel or capstan as shown in Figure 10. The resulting size of the fiber relative to the size of the preform is determined by the pulling speed of the fiber relative to the feed rate of the preform.

$$\text{Draw Speed} = \text{Feed Speed} \ (D^2/d^2)$$

where D = preform diameter and d = fiber diameter.

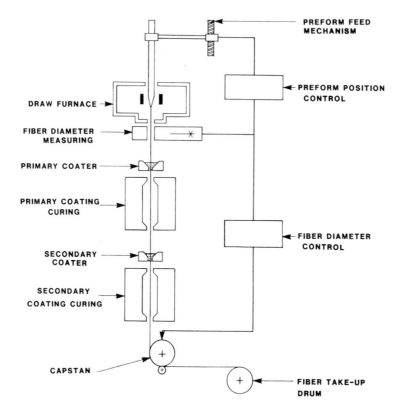

FIGURE 10.   Fiber drawing apparatus.

All high grade optical waveguide fibers are essentially pure silica. Consequently, it is necessary to heat the drawing end (root) of the preform to at least 2000°C to soften the glass enough to readily stretch it. Thus, the first problem is a heat source capable of high temperature. More specifically, the problem is the material of the heat source. A temperature of 2000°C is beyond the melting and vaporization temperature of most materials.

The second problem is the size and needed accuracy of the fiber. Most fibers produced have a diameter of 125 μm (less than 0.005 in.) with deviations of less than 1 μm required for cabling and splicing. This requires a method of continuously measuring the fiber at high speeds to submicron accuracies and a responsive feedback system to keep the preform feed and fiber draw speeds in the correct ratio.

The third problem in fiber drawing is fiber strength. Conventional glass products have a use strength of 7 to 20 MPa (1000 to 3000 psi), but optical waveguide fibers must have guaranteed strength of at least 350 to 700 MPa (50,000 to 100,000 psi) to survive cabling processes and long-term use conditions. Since the potential strength of silica glass is near 1,000,000 psi, the required strengths are possible. It is submicroscopic damage to the surface of the fiber, as well as slight imperfections and particulate contaminants on the surface, that reduce the strength. With great care, these can be eliminated, or at least minimized.

In addition to possible surface contamination, a fourth problem is possible contamination of the core glass itself and resulting degradation of optical properties. This is possible due to the mobility and ready diffusion of most elements in the glass at elevated temperatures. Consequently, any volatilization of iron, copper, nickel, etc. from the best source can be readily absorbed on the preform and diffused into the core glass. Contaminants of only a few ppb can result in significantly increased attenuation.

The drawing of optical waveguide fibers requires that the root of the preform be heated from 1900 to 2200°C. Three methods are used: a flame, a high power laser, and various types of electric furnaces.

A flame heat source must be near a stoichiometric H_2 and O_2 torch to achieve the required temperature. The preform must be heated uniformly around its circumference to produce round fiber. Consequently, the burner assembly must be a precise annular design or a multitude of small identical torches. Because of the heat of the flame and the radiation from the hot preform, the burner assembly must be thoroughly water cooled to prevent it from being overheated and destroyed. The most severe limitations for a flame draw are the convective currents and turbulence caused by the flame itself and the products of combustion that disturb the small delicate fiber and make it difficult to maintain accurate fiber dimensions.

When first proposed, laser heating of preforms for fiber drawing seemed ideal.[22] A beam from a high powered CO_2 laser can be focused to produce extremely high temperatures. The wavelength of the radiation (10.6 μm) is readily absorbed by the silica glass, and there are no combustion products to produce turbulence. A complex optical system is needed to transform the linear radiant beam into a symmetrical annular source, but several have been devised. Fibers are now successfully drawn with laser heating, but the multi-kilowatt lasers that are needed are state-of-the-art. They are unreliable in a manufacturing environment and expensive.

An electric furnace has always been the most popular heating method in the common drawing process. However, in the conventional draw process, the operating temperatures are only 900 to 1700°C, and the heat source can be resistance wire elements made of nickel-chrome alloys or platinum alloys wrapped around a ceramic tube. The temperatures required for silica drawing are beyond the melting temperature of those materials.

Electric resistance furnaces are often used for waveguide drawing.[23-24] The resistance elements in these furances have been changed to thermally resistant materials such as tungsten, molybdenum, or graphite. However, these materials cannot be exposed to oxygen when hot or they will burn. This necessitates keeping the heating elements in a vacuum, or more commonly, an inert gas atmosphere.

A difficulty that occurs with the resistance furnace is the inability to connect the power leads to the high temperature resistance elements. It is again a materials temperature problem and is usually solved with elaborate water cooling techniques. One alternative is to couple the electrical power into the heating via an induction coil.[25] This eliminates the problem with the power leads.

One very unique solution to all of the problems is the zirconia induction furnace. In this design, the same zirconia tube is used as the furnace tube and as the heating element. This is possible because the ceramic zirconia material becomes electrically conductive above 1000°C and can be heated in a high frequency induction field. Most importantly, it does not require any protective atmospheres. This is, at the present, the most widely used equipment for the drawing of optical waveguide fiber.

The methods of fiber splicing and connectorization require precise control of the fiber diameter. Also, excessive diameter variation can produce mode coupling and resulting optical attenuation.

The most important factor in precise diameter control is a quiescent atmosphere at the preform root and neck-down region. Any turbulence or excessive convection will result in high frequency diameter variations. The solution is to create a seal or, at worst, close clearances where the preform enters the furnace to prevent excessive vertical flow through the furnace due to the chimney effect. Sometimes, an iris or some other closure is used at the bottom of the furnace where the fiber exits to also minimize the flow. In the case of exposed element furnaces, the purge gas system must be carefully designed to avoid the turbulence and excess flow conditions. Another related condition is convection within the furnace. If the interior volume has temperature gradients that are too large, convection currents will be generated within the furnace that will cause diameter variations. The fiber, while being drawn, is very fluid and is disturbed by even minute perturbations. The draw tower assembly must be rigid and decoupled from the building structure to prevent mechanical vibrations from being transmitted into the preform or fiber. Similarly, the downfeed mechanism must be smooth acting. Any high frequency perturbations in downfeed speed will be greatly amplified in diameter variations. The pulling capstan or pinch wheel must also be precise. Any eccentricity of the pulling mechanism or erratic drive motor characteristics result in immediate diameter variations.

Given a quiescent atmosphere and absence of mechanical perturbations, the actual diameter control requires a measurement system and a closed loop control system to the pulling mechanism. The fiber diameter is measured with two types of laser systems. One system uses the interference pattern generated by the fiber to infer its diameter.[26] The other system uses a scanning laser beam and fiber shadow to determine the diameter. Both systems are capable of resolution to a fraction of a micron. Given the diameter measurement, a control system determines the error and then adjusts the pulling mechanism speed either faster or slower to make the fiber smaller or larger. The control system is somewhat complicated because there is a time lag between when the fiber is formed within the furnace and when it can be measured near the furnace exit. Another complication is that the required response is not linear, i.e., the speed change must be a square of the diameter error. In spite of these complications, a frequency response of 30 Hz for the combined measurement, control, and pulling mechanism system is adequate.

The greatest difficulty in creating strong fibers is the requirement of not touching the fiber with anything until a durable coating has been applied to protect the surface. Applying the protective coating uniformly and concentrically *without* touching the fiber is very difficult. Many different techniques are used with varying degrees of success.

Being able to protect the surface from damage, the requirement is then to produce an inherently strong fiber. To this end, clean room conditions are maintained throughout the preform making and fiber drawing environments. Also, high draw temperatures are employed to heal any defects in the preforms and to burn off, volatize, and dissolve any contaminants.

IV. OPTICAL MEASUREMENT

A. Optical Characterization

The optical characterization of multimode optical fibers can be grouped roughly into two categories. The first category is quality control measurements. These measurements determine the grade and selling price of manufactured fibers. The measurements must be fast, reliable, and require a minimum of technical skill and knowledge to be performed.

The other category of optical measurements is diagnostic measurements. These are utilized by manufacturers to optimize their process for a specific product type, develop new products, and diagnose manufacturing problems. In this chapter, both quality control and diagnostic measurements for multimode fibers will be discussed. Techniques for characterizing single-mode fibers are still being developed and will not be presented.

B. Quality Control Measurements

The standard optical parameters that are measured in quality control are attenuation and bandwidth, both at several wavelengths and NA. Improvements in fiber quality have made tremendous demands on measurement equipment. In a production environment, measurement uncertainty is typically 0.1 dB/km for attenuation and 5 to 10% for bandwidth measurements. In this section the quality control measurements of attenuation, bandwidth, and NA will be discussed.

C. Attenuation Measurements

The attenuation of an optical fiber depends on the launch conditions that are used to couple the optical source to the fiber. There is general agreement that restricted launch conditions should be used. There are two basically equivalent techniques that are currently being proposed by the EIA[27] for obtaining a restricted launch. These are commonly referred to as the beam optics and the mode filter techniques. Examples of mode filters are dummy fibers and mandrel wraps. In this section only the mandrel wrap filter will be considered since it is the technique most commonly used.

In the beam optics technique, the optics are adjusted so as to produce an image launch spot having a diameter at the 5% intensity points equal to 70% of the nominal core diameter and a launch NA equal to 70% of the nominal fiber NA. The launch beam must be imaged and centered on the fiber core to within a tolerance of 5 μm.

Use of the mandrel wrap mode filter first requires that the test fiber be overfilled in both launch NA and spot size. The input fiber end is then subjected to a helical wrap within a 15 mm region of a smooth mandrel, with no overlapping turns and no tension beyond that required to maintain contact between the fiber and mandrel. The diameter of the mandrel and the number of turns are chosen to produce a far-field pattern similar to that produced using the beam optics launch. Attenuation measurements using this type of launch show excellent agreement with measurements done using the beam optics launch for most fibers.[28]

D. Cut-Back Technique

The cut-back technique provides the most precise measurement of optical attenuation. Though the technique is destructive, the benefits far outweigh this drawback.

The cut-back technique is performed by first measuring the optical power from the end of the fiber being tested using appropriate launch conditions. Then, without disturbing the launch, the fiber is broken approximately 2 m from the launch end, and the power is measured at that point. The attenuation of the fiber is then calculated as:

$$loss = -\frac{10}{L} \log \frac{P_1}{P_2}$$

where L = fiber length in km, P_1 = power from the long length, and P_2 = power from the short length.

A typical measurement apparatus used for the cut-back method is shown in Figure 11. The components required to obtain restricted launch conditions are also shown in the diagram.

E. Bandwidth Measurements

Bandwidth measurements also depend on launch conditions. The launch conditions that are normally used are obtained by using a mode scrambler. This consists of a concatenation of three 2-m pieces of fiber, with the two end pieces being of a graded index design and the center piece being a step-index fiber. The optical source is coupled to one end of the scrambler, and the other end is imaged onto the end of the test fiber. The effect of the mode scrambler is to distribute the optical power from the source over the spatial and angular extent of the test fiber. Using these launch conditions, the repeatability of bandwidth measurements is greatly improved as compared to the precision obtained when the optical source is imaged directly onto the test fiber.

Bandwidth measurements can be performed in both the time and frequency domains. Both domains have their own advantages. Due to the availability of pulsed laser diodes before CW laser diodes, most quality control facilities began doing the measurement in the time domain and, as a result, are still using this technique. In this section the two measurements techniques will be described and the advantages of each will be discussed.

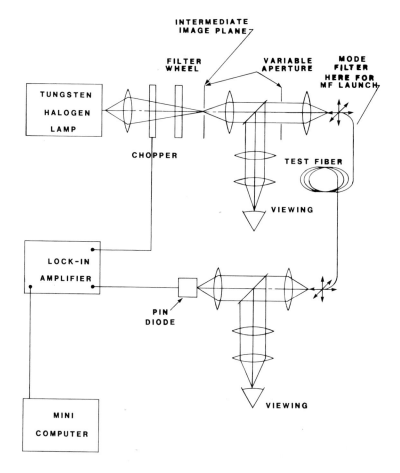

FIGURE 11.   Cut-back attenuation measurement.

1. Time Domain

Time domain bandwidth measurements are performed by launching a narrow optical pulse from a pulsed or mode-locked laser into the fiber and detecting the broadened pulse at the output of the fiber. The input and output pulses are digitized using a sampling oscilloscope, and their Fourier Transforms are obtained using a Fast Fourier Transform (FFT) algorithm on a minicomputer.

The response of the electrical equipment is then subtracted from the total response to get the response of the fiber. The biggest drawback, of course, is that a minicomputer is needed to obtain the FFT. This technique also has a limited dynamic range as compared to the frequency domain measurement. One of the main advantages is the measurement system bandwidth that can be achieved using this technique. Typically the detected pulse is limited in time by the response of the detector being used. In most cases, this response time is on the order of 500 psec, which gives a system bandwidth on the order of 1 GHz. It is very difficult to obtain this kind of bandwidth in the frequency domain where laser modulation circuits with flat frequency responses out to 1 GHz are required. A block diagram of a typical time domain measurement is shown in Figure 12.

2. Frequency Domain

The swept frequency measurement of optical fiber bandwidth is very similar to the standard swept frequency technique utilized to measure the frequency response of electrical equipment. The measurement consists of a sinusoidally modulated light source whose frequency is swept

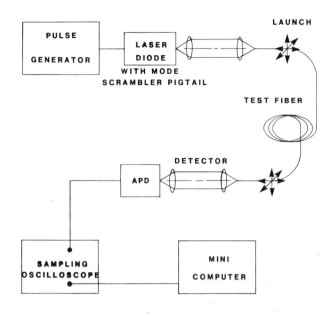

FIGURE 12. Time domain bandwidth measurement.

from DC to approximately 1 GHz. The light is coupled into the fiber being tested and then detected at the output end of the fiber using a high speed photodetection circuit. The output of this circuit is connected to a spectrum analyzer that displays the combined frequency response of the fiber and the measurement equipment. To obtain the frequency response of the fiber, the response of the measurement equipment is determined using a short length of fiber between the source and receiver. This response is then subtracted from the total response that was measured for the long length of fiber.

The main disadvantage of this method is the requirement that the modulation and detection circuits must have virtually flat frequency responses from DC to 1 GHz. This can be achieved; however, noise effects and laser instabilities still limit the maximum bandwidth that can accurately be measured using this technique. A block diagram of a typical frequency domain measurement is shown in Figure 13.

F. Numerical Aperture (NA) Measurement

The NA measurement is performed by sweeping a detector over an arc that is located a distance from the end of the fiber. The distance must be sufficiently large to insure that it is in the far field of the fiber. An intensity pattern as a function of angle is then obtained. The definition of NA that is most commonly used is the sine of the angle between the axis of the fiber and the point in the far-field intensity pattern at which the intensity has fallen to 5% of its own axis value. This technique is quite simple and does not require a large amount of expensive equipment. A diagram of a typical far-field measurement system is shown in Figure 14.

G. Diagnostic Measurements

Diagnostic measurements are used in the optimization of fiber fabrication. The most commonly used measurements are those that provide information on the attenuation mechanisms in the fiber and the sources of pulse broadening that limit the bandwidth of the fiber. In addition, measurements of the refractive index profile are also done as a diagnostic tool

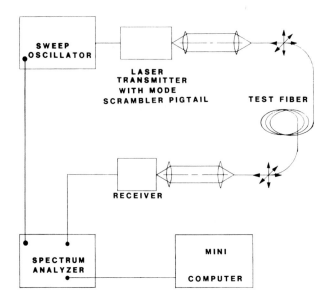

FIGURE 13. Frequency domain bandwidth measurement.

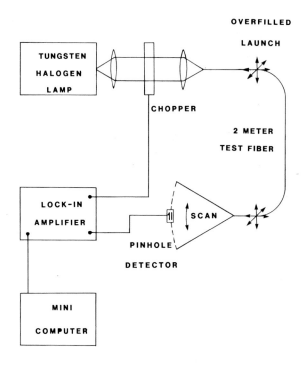

FIGURE 14. Numerical aperture measurement.

to detect any deviations from an ideal profile that may have occurred during fabrication. In this section, some of the commonly used diagnostic measurements will be described.

H. Differential Mode Delay (DMD)

Differential mode delay measurements[29] are extremely useful for optimizing the bandwidth of multimode graded-index fibers. In essence, the measurement determines the differences in transit time between the different modes in the fiber. This information provides a means of determining how the refractive index profile should be modified to equalize the transit times of the different modes.

The technique most commonly used to excite the different modes is to focus a small spot of light on the end face of the fiber and vary its radial position. The relation between the compound mode number, m; position, r; and angle, Θ; of a launch ray for a fiber with a parabolic index profile

$$m/M = (r/a)^2 + (\sin \theta / \sin \theta_c)^2$$

where M is the total number of modes, a is the core radius, and Θ_c is the critical angle. Therefore, for a beam launched parallel to the fiber axis at a radial distance, r, from the fiber axis, the principal mode number is given approximately by

$$m/M = (r/a)^2$$

From this relation it can be seen that by varying the position of the spot on the end of the fiber, different mode groups can be excited. The difference in mode transit times is obtained by launching a very narrow optical pulse from a mode locked or pulsed laser and detecting its arrival time at the output of the fiber using a fast detector and a sampling oscilloscope. Since absolute transit times are not of interest, the arrival time of the lowest order mode group is normally assigned an arrival time of zero, and subsequent mode arrival times are measured with respect to it.

Once the differential transit times of the different mode groups are known, the defects in the refractive index profile that caused unequal mode transit time can be estimated and corrected in subsequent fibers. Normally several iterations are required before an ideal profile shape is obtained. In addition, by doing the measurement at several different wavelengths, data can be obtained that can be used to optimize the fiber bandwidth for operation at an arbitrary wavelength of interest. The apparatus for doing DMD measurements is shown in Figure 15.

I. Differential Mode Attenuation (DMA)

Differential mode attenuation[30] is basically the attenuation analog to DMD. The objective of the measurement is to determine the attenuations of the different mode groups in a multimode fiber. By determining the attenuation as a function of mode group, insight into attenuation causing mechanisms can be gained. For example, by comparing DMA measurements done at 1.06 and 1.39 μm, an estimate of the radial distribution of OH^- ion contamination can be obtained since the OH^- molecule has a resonance at 1.39 μm and will therefore cause increased attenuation at this wavelength.

The DMA measurement is normally performed using the same type of mode excitation used in DMD measurements in conjunction with the cut-back technique for determining attenuation. First, the optical power from the fiber under test is recorded as a function of launch spot position. The fiber is then broken 2 m from the launch end, and the procedure is repeated. The differential attenuation is then calculated using:

$$A(r) = -\frac{10}{L} LOG[P_1(r)/P_2(r)]$$

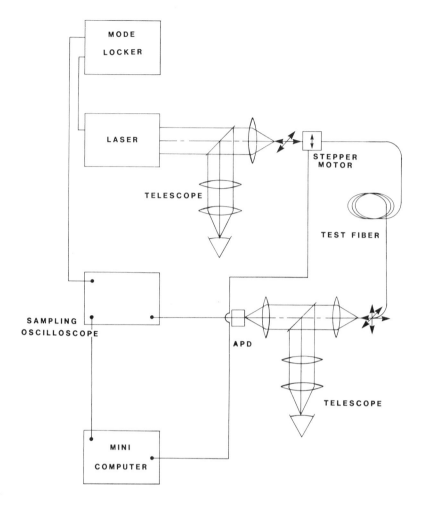

FIGURE 15. DMD measurement.

where r is the launch position of the spot, $P_1(r)$ and $P_2(r)$ are the long and short power levels as a function of launch position, and L is the length of the fiber in kilometers. To produce a true differential measurement, the attenuations can be normalized to the m = O mode group. However, the technique also produces an accurate measure of absolute attenuation for each of the mode groups. A typical measurement system for DMA is shown in Figure 16.

J. Bandwidth Spectra Measurements

The bandwidth of a multimode fiber is a function of wavelength due to the differences in dispersion between the core and cladding glasses of the fiber. Much information about the composition and design of a fiber can be obtained by measuring bandwidth as a function of wavelength. A source of tunable optical radiation to measure bandwidth can be obtained from a dye laser for the 0.7 to 0.85 μm region, and a single pass Raman laser can be used to cover the 1.1 to 1.7 μm region. An example of a bandwidth spectrum plot is shown in Figure 17.

K. Refractive Index Measurements

The refractive index profile of a multimode fiber is the single most important parameter

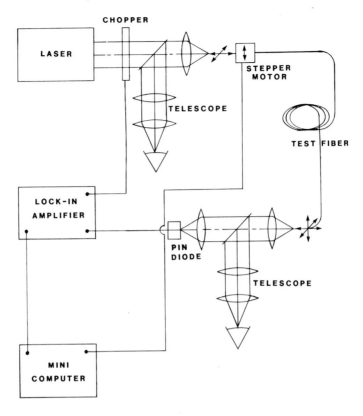

FIGURE 16. DMA measurement.

that determines the propagation characteristics of the fiber. Because of its importance, several techniques have been developed to measure the refractive index profiles of both fibers and preforms.

L. Preform Index Profiling (PIP)

The refractive index profile of a preform can be reconstructed from data on the refraction of rays passing transversely through the preform as a function of distance from the preform center.[31] Several techniques exist[32-33] to measure this so-called "deflection function". The details of these techniques will not be presented here.

Due to the large size of a preform, extremely high spatial resolution refractive index measurements can be obtained. A typical example of the refractive index profile of a multimode preform is shown in Figure 18. As can be seen, the measurement resolves virtually each deposition layer in the core.

Inherent in the computation of refractive index from the deflection function is the assumption that the preform is circularly symmetric. Calculations for an elliptical parabolic index preform show that the maximum error in the refractive index difference is given by

$$\left(\frac{\delta\Delta}{\Delta} = \pm \frac{a - b}{a}\right)$$

where a and b are the major and minor axis dimensions.

M. Fiber Refractive Index Measurements

One direct method to measure the refractive index profiles of fibers is the refracted near-

BANDWIDTH SPECTRUM

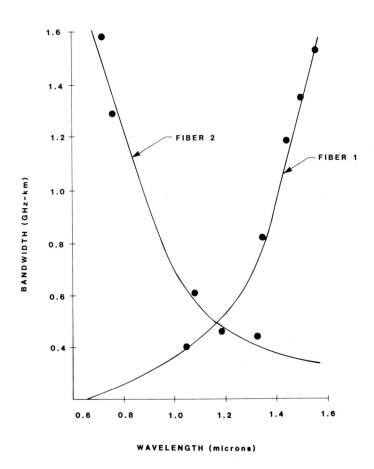

FIGURE 17.   Bandwidth spectrum.

field technique.[34] This measurement typically has spatial resolution of less than 1 μm and a resolution in refractive index of less than 0.0002.

The refracted near-field technique consists of focusing a laser beam onto the end of the test fiber and recording the refracted power above a certain angle defined by a disc placed behind the fiber. The power is recorded as a function of spot position. Using a linear calibration factor, the refractive index profile of the fiber can be obtained directly from the refracted power. The main disadvantage of this technique is that the fiber must be cleaved perpendicular to the axis of the fiber, and the end face must be flat and contamination free. The advantage of this technique is that a computer algorithm is not required to determine the refractive index from the measured data. For this reason, the technique is fast, simple, and requires a minimum of hardware.

Due to the limitations on the size of this section, a detailed description of the measurements was not possible. For additional information on optical fiber measurements, it is suggested that the reader consult Reference 35.

V. MECHANICAL PROPERTIES

Although mechanical reliability is not the most important criterion for lightguide manufacturing, it is a critical constraint and, therefore, worthy of discussion. In order for wave-

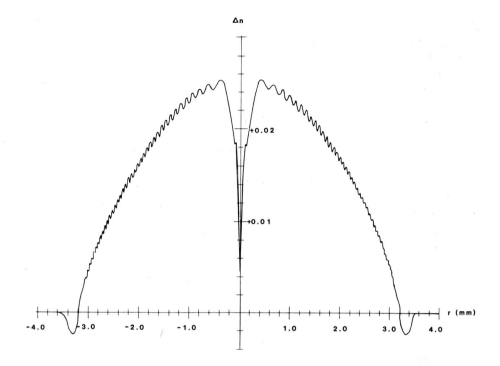

FIGURE 18.   Preform refractive index profile.

guide manufacturing to be competitive with conventional guides and for these cables to be accepted in the marketplace, the fibers must maintain mechanical integrity throughout their manufacture, installation, and service. Installation stresses can be quite high with stresses of 175 MPa (25 kpsi.) not uncommon for pulling cable through troughs and ducts. To insure mechanical integrity, the manufacturer must test to assure the minimum strength on each fiber. To complicate this quality assurance process, time dependent failure can cause fiber to fail at loads much lower than the assured minimum strength under certain environmental and loading considerations. The purpose of this section is to explain some of the basics in providing fiber that will maintain integrity throughout a reasonable service life.

A. Tensile Strength of Fiber

The theoretical strength of any material is ultimately determined by the cohesive strength of the constituent atoms. This strength can be estimated by the materials elastic modulus.[36] In silicate glasss, theoretical strength has been found as high as 14 GPa (2000 kpsi.).[37] The flaws serve as stress intensifiers at which strain energy is released in the form of crack tip extension. Failure stress is related as such:

$$\sigma = \sqrt{\frac{2\epsilon\gamma}{\pi a}}$$

where ϵ = Young's modulus, γ = surface energy, σ = failure stress, and a = half crack length.

If a fiber is brought under load, the largest flaw within the fiber will control the strength of that fiber just as a weak link fails in a chain. Therefore, the control of fiber strength is dependent on the control of the flaws in the glass.

Fiber strength is dependent on several factors. Glass quality has the largest effect on strength. For high strength fiber, both the bulk and surface glass must be the best obtainable.

Synthetic silica (SiO_2) made from the flame hydrolysis of $SiCl_4$ generally provides glass free from particulate and inclusions, but is costly. Natural silica is cheaper, but is not impurity free. Economics demands which is to be used.

Cleanliness also has a large effect on the strength of the fiber. At each step of the manufacturing process, care must be taken to keep contamination at a minimum. Refractory particles such as those used in torch and furnace materials are especially unwanted since they provide gross flaws as shown in Figure 19.

Mechanical damage is the last important factor deleterious to fiber strength. In the redrawing process, care must be taken to keep the fresh or pristine glass surface unabraded until it is coated with an appropriate coating.

Although it is widely known which factors affect fiber strength, it is difficult to eliminate all the flaws because of their size and frequency. Since the glass has a distribution of flaw sizes, the strength of the fiber will be a statistical quantity. Because of this, it is necessary to discuss glass strength in terms of probability. An accepted method for describing the strength probability dependence is the Weibull function:[39]

$$\ln\ln \left(\frac{1}{1 - F} \right) = \ln L + m \ln \sigma$$

where F = failure probability, L = length, volume, or diameter, and σ = stress or strength.

Figures 20 and 21 show two such Weibull plots of typical production fiber for two different gauge lengths. It can be seen in Figure 20 that the probability in Figure 20 of finding a flaw which causes failure below 4 GPa ($\simeq$500 kpsi.) is extremely low while in Figure 21 it is 40%. This is due to the longer specimen length in Figure 21.

It can be seen in both Figures 20 and 21 that no failures occurred below 700 MPa (100 kpsi.). These fibers have been screened to eliminate any flaws of the critical size to cause failure at this level. This enables the manufacturer to guarantee a minimum strength, in this case 100 kpsi. This process, termed prooftesting, is done to all fiber by loading the fiber to a desired stress for a brief period (<1 sec). Any flaws of critical size cause fiber breakage.

B. Lifetime Forecasting

The service life of lightguides is a basic concern. Manufacturers must guarantee a minimum lifetime for their fiber so customers will deem them a reliable medium. Time-dependent failure has been studied extensively by many researchers. It has been found that the fiber fatigues under the static loads seen in service (<10 kpsi.) by subcritical crack growth.[40] The lifetimes are greatly affected by environmental factors such as temperature and humidity.

The minimum fiber strength or proof stress also affects the lifetime. Higher proof stress increases the minimum guaranteed lifetime of the prooftested fiber. Figure 22 predicts fiber lifetime for ambient conditions under various proof stress. Although the lifetime is greatly increased by raising the proof stress, this increases fiber breakage. The proof stress must be selected to yield both an adequate lifetime and an acceptable fiber yield. For example, for a given proof stress of 345 MPa (50 kpsi.) and a service stress of 125 MPa (18 kpsi.) the lifetime prediction will be 100 years. In general, for silica fibers the proof stress should be three times the service stress.

Researchers are now working on various ways to seal the fiber hermetically from the environment. This will greatly enhance the fiber's resistance to static fatigue. Two approaches to hermetic coatings have been aluminum coatings and silica nitride coatings.[41-42] Although both approaches have shown some success, neither has been entirely successful. The aluminum coatings show some microbending problems while the nitride coatings severely lower the mean strength of the fiber.

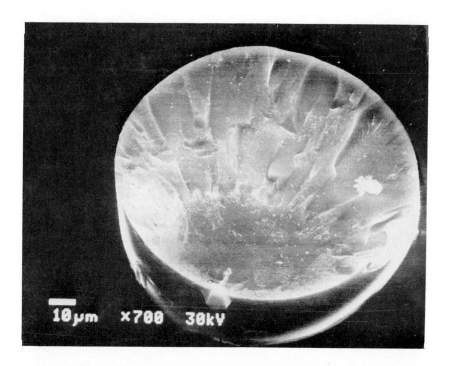

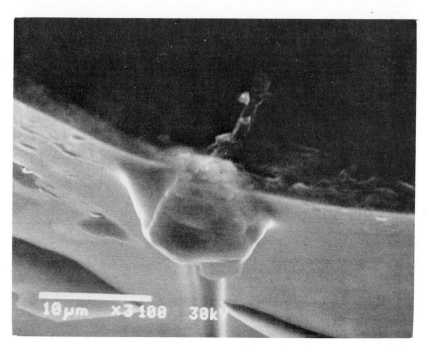

FIGURE 19. Scanning electron micrographs of refractory particle embedded in fiber surface causing fiber failure under proof stress.

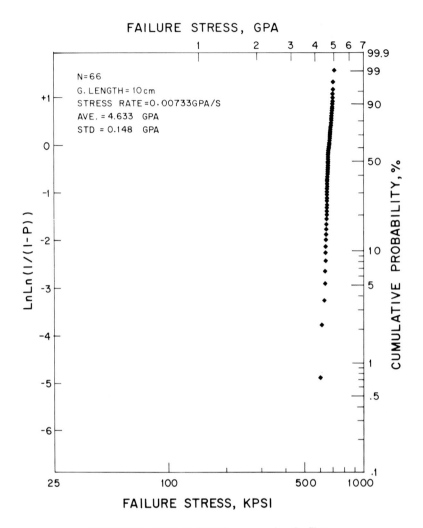

FIGURE 20.   Weibull plot, 10-cm gauge length fiber.

ACKNOWLEDGMENTS

The author is greatly indebted to his colleagues Dr. Peter H. Prideaux, John E. Chapman, Thomas P. Leo, Dr. William C. Meixner, and Carl D. Andrysiak at Valtec for their contributions to this chapter. He is also thankful to Joseph DePasquale, Shirley Bayer, Patricia Forti, Karen Nickerson, and Patrick McCormick for their kind help in reviewing, typing, and drafting.

PROBLEMS

1.   A waveguide preform manufactured by the MCVD process is measured to be 70 cm long and 19 mm in diameter. If it is drawn to 125 µm, what will be the fiber length assuming 90% efficiency? How long will this take at 1 m/sec and at 5 m/sec?

2.   Discuss why it is more difficult to draw silica fibers compared with fiber for more conventional applications (fiber with lower melting temperatures).

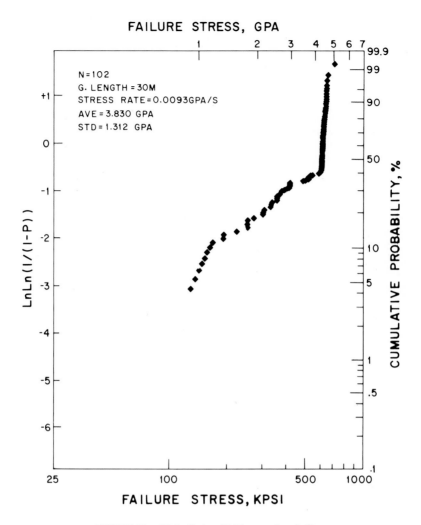

FIGURE 21. Weibull plot, 30-M gauge length fiber.

3. Explain how the zirconia element in a high frequency induction furnace is heated. What is the mechanism of heating through induction?
4. Explain why it is important to have waveguide fiber dimensional control.
5. Glass fibers 125 μm in diameter are tensile loaded in an Instron machine to 500 g. If the fibers fractured during the test, what size Griffith flaws would be expected? (E = 70 GPa, γ = 4000 ergs/cm^2)
6. The mirror, a flat smooth region observed adjacent to the fracture origin in low fracture stress specimens of brittle materials, can be empirically related to failure stress by σ = 2.4/r where r = mirror radius. If fiber was prooftested to 700 mn/m^2, what would be the minimum mirror radius observed? Flaw size observed?
7. Explain the fiber's strength dependence on sample length illustrated in Figures 20 and 21.
8. Estimate the proof stress necessary to insure a minimum lifetime of 20 years at a service load of 500 g (fiber diameter = 125 μm).

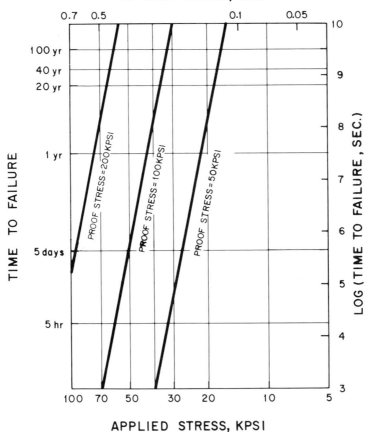

FIGURE 22.   Applied stress (in service), kpsi.

REFERENCES

1. **Berkey, G. E. and Sarkar, A.,** Single mode fibers by the OVD process, paper ThCC5 presented at Fifth Topical Meeting on Fiber Commun., Phoenix, Ariz., 1982.
2. **Murata, H. and Inagaki, N.,** Low-loss single-mode fiber development and splicing research in Japan, *IEEE J. of Quantum Electron.,* QE-17, 835, 1981.
3. **Olshansky, R. and Keck, D. B.,** Pulse broadening in graded-index optical fibers, *Appl. Opt.,* 15, 483, 1976.
4. **Gambling, W. A., Matsumara, H., and Ragdale, C. M.,** Zero mode dispersion in single-mode fibers, *Electron. Lett.,* 14, 618, 1978.
5. **Irven, J., Byron, K. C., and Cannell, G. J.,** Dispersion characteristics of practical single mode fibers, presented at 75th Eur. Conf. on Opt. Commun., Copenhagen, Denmark, 1981.
6. **Schultz, P. C.,** Optical absorption of the transition elements in vitreous silica, *J. Am. Ceram. Soc.,* 57, 309, 1974.
7. **Kapany, N. S.,** *Fiber Optics,* Academic Press, New York, 1967, 429.
8. **Pearson, A. D. and French, W. G.,** Bell Labs Rec. 50-103-9, 1972.

9. **Newns, G., Beales, K. J., and Day, C. R.,** Development of low-loss optical fibres, Proc. Int. Conf. on Int. Opt. & Opt. Fiber Commun., Tokyo, 1977, 609.

10. **Tanaka, S., Inada, K., Akimoto, T., and Kojima, M.,** Silicone clad fused silica core fiber, *Electron. Lett.,* 11, 153, 1975.

11. **Litovitz, T. A. and Macedo, P. B.,** Catholic University of America, U.S. Pat. Pend.

12. **Keck, D. B. and Schultz, P. C.,** U.S. Patent 3,711,262, 1973.

13. **Macedo, P. B., Simmons, J. H., Olson, T., Mohr, P. K., Samanta, M., Gupta, P. K., and Litowitz, T. A.,** Molecular stuffing of phasil glasses for graded index optical fibres, Proc. 2nd Eur. Conf. Opt. Fiber Commun., Paris, 1975, 37.

14. **MacChesney, J. R., et al.,** *Am. Ceram. Soc. Bull.,* 52 (Abstr.), 704, 1973.

15. **Izawa, T., Kobayashi, S., Sudo, S., and Hanawa, F.,** Continuous fabrication of high silica fibre preform, Proc. Int. Conf. Int. Opt. & Opt. Fiber Commun., Tokyo, 1977, 375.

16. **Geittner, P., Kyppers, D., and Lydtin, H.,** Low-loss optical fibres prepared by plasma-activated chemical vapor deposition (CVD), *Appl. Phys. Lett.,* 28, 645, 1976.

17. **Kuppers, D., Koenings, J., and Wilson, H.,** Code-position of glassy silica and germania inside a tube by plasma-activated CVD, *J. Electrochem. Soc.,* 123, 1079, 1976.

14. **MacChesney, J. B., O'Conner, P. B., Simpson, J. R., and DiMarcello, F. U.,** Multimode optical waveguides having a vapor deposited core of germania doped borosilicate glass, *Am. Ceram. Soc. Bull.,* 52, 704, 1973 (abstract).

18. **Kuppers, D., Koenings, J., and Wilson, H.,** Deposition of fluorine doped silica layers from a $SiCl_4/SiF_4/O_2$ gas mixture by the plasma-CVD method, *J. Electrochem. Soc.,* 125, 1298, 1978.

19. **Fleming, J. W. and O'Connor, P. B.,** High rate lightguide fabrication technique, in *Phys. Fiber Opt. Adv. Ceram.,* Vol. 2, Bendow, B. and Mitra, S. S., Eds., American Ceramic Society, Columbus, Ohio, 1981, 21.

20. **Fleming, J. W.,** Progress in plasma prep. of lightguides, presented at the 23rd Int. Congr. Glass, Albuquerque, N.M., Poster Session, July 6 to 11, 1980.

21. **Fleming, J. W. and Raju, V. R.,** Low optical attenuation fibers prepared by plasma enhanced MCVD, in paper WD2, Tech. Dig. 3rd Int. Conf. Int. Opt., Opt. Fiber Commun., paper WD2, San Francisco, Calif., April 27 to 29, 1981.

22. **Jaeger, R. E.,** Laser drawing of glass fiber optical waveguides, *Am. Cer. Soc. Bull.,* 55, 270, 1976.

23. **Payne, D. N. and Gambling, W. A.,** A resistance heated high temperature furnace for drawing silica based fiber for optical communications, *Am. Cer. Soc. Bull.,* 55, 195, 1976.

24. **DiMarcello, F. V. and Williams, J. C.,** Reproducibility of optical fibers prepared by a CVD process, *Inst. Elect. Eng.,* 132, 36, 1975.

25. **Runk, R. B.,** A zirconia induction furnace for drawing precision silica waveguides, 2nd Tech. Dig. Top. Meet. Opt. Fiber Transmission, paper TuB5, Williamsburg, Va., February 22 to 24, 1977.

26. **Smithgall, D. H., Watkins, L. S., and Frazee, R. E., Jr.,** High speed noncontact fiber diameter measurement using forward scattering, *Appl. Opt.,* 16, 2395, 1977.

27. FOTP #50, Light launch conditions for long length graded-index optical fiber spectral attenuation measurements, Mil Spec 1678, 1981.

28. **Franzen, D. L., Day, G. W., Danielson, B. L., Chamberlain, G. E., and Kim, E. M.,** Interlaboratory measurement comparison to determine the attenuation and bandwidth of graded-index optical fibers, *Appl. Opt.,* 20, 2412, 1981.

29. **Jeunhomme, L. and Pocholle, J. P.,** Selective mode excitation of graded index fibers, *Appl. Opt.,* 17, 463, 1978.

30. **Olshansky, R. and Oaks, S. M.,** Differential mode attenuation measurements in graded-index fibers, *Appl. Opt.,* 12, 1930, 1978.

31. **Chu, P. L.,** Nondestructive measurement of index profile of an optical fiber-preform, *Electron. Lett.,* 13, 736, 1977.

32. **Sasaki, I., Payne, D. N., Mansfield, R. J., and Adams, M. J.,** Variation of refractive-index profiles in single-mode fibre preforms measured using an improved high-resolution spatial filtering technique, Proc. 6th Eur. Conf. on Opt. Fiber Commun., York, U.K., September 16 to 19, 1980, 140.

33. **Chu, P. L. and Whilbread, T.,** Measurement of refractive-index profile of optical-fibre preform, *Electron. Lett.,* 13, 59, 1979.

34. **Stewart, W. J.,** A new technique for measuring the refractive index profiles of graded optical fibers, paper C2-2, Tech. Dig. Int. Conf. Int. Opt., Opt. Fiber Commun., Tokyo, 1977.

35. **Marcuse, D.,** *Principles of Optical Fiber Measurement,* Academic Press, New York, 1981.

36. **Irwin, G. R. and Wells, A. A.,** A continuum mechanics view of crack propagation, *Met. Rev.,* 10, 223, 1965.

37. **Proctor, B. A., Whitney, I., and Johnson, J. W.,** The strength of fused silica, *Proc. R. Soc. London,* Ser. A297, 534, 1967.

38. **Griffith, A. A.,** The phenomena of rupture and flow in solids, *Philos. Trans. R. Soc. London,* Ser. A221, 163, 1920.
39. **Weibull, W.,** A statistical distribution function of wide applicability, *J. of Appl. Mech.,* 18, 293, 1951.
40. **Charles, R. J. and Hillig, W. B.,** The kinetics of glass failure by stress corrosion, Symp. Mechanical Strength of Glass and Ways of Improving It, Union Scientifique Continentale du Verre, 1961, 511.
41. **Wysocki, J. A., Blair, O. R., and Robertson, E. D.,** Long term strength of metal coated fibers, in *Physics of Fiber Optics, Advances in Ceramics,* Vol. 2, Bendow, B. and Mitra, S. S., Eds., American Ceramic Society, Columbus, Ohio, 1981, 134.
42. **Hiskes, R.,** Improved fatigue resistance of high strength optical fibers, paper WF4, in Tech. Dig. Top. Meet. Opt. Fiber Commun., 1979, Washington, D.C., March 6 to 8, 1979.

Chapter 3

PROPAGATION IN OPTICAL FIBERS

Robert G. Olsen and David A. Rogers

TABLE OF CONTENTS

I. INTRODUCTION

The subject of this chapter is the attenuation and distortion of pulses which propagate in monomode and multimode optical fibers. For monomode fibers it is prudent to use modal theory to describe the propagation of light pulses. Hence this is the approach used here. The use of modal theory requires an understanding of solutions to the wave equation in cylindrical cordinates (Bessel functions). This process requires a substantial mathematical background. For multimode fibers, on the other hand, simple ray optics can be used to describe essential propagation characteristics. This approach, which is used here, has the advantage of being less mathematical and more physically intuitive than modal theory. For this reason some readers may choose to study optical fiber propagation by reading Sections II, III, and IV followed by VII through to the end. This approach will provide the reader with an understanding of many of the fundamental problems in optical fiber propagation without requiring advanced mathematical skills.

II. MODES IN STEP-INDEX FIBERS

A thorough understanding of optical fiber propagation requires an appreciation of mode theory. Based on Maxwell's equations, this approach permits a comprehensive vision of propagation in step-index fibers.

It is well known[1,2] that applying Maxwell's equations to a linear, homogeneous isotropic region leads to the wave equation in matrix form

$$\nabla^2 \begin{bmatrix} \overline{E} \\ \overline{H} \end{bmatrix} = \mu\epsilon \frac{\partial^2}{\partial t^2} \begin{bmatrix} \overline{E} \\ \overline{H} \end{bmatrix} \tag{1}$$

where $\overline{E}$ and $\overline{H}$ are the electric and magnetic field intensities, μ is the material permeability, and ϵ is the permittivity of the medium. In fibers, μ is usually μ_o, the permeability of free space.

In scalar form, Equation 1 is written as

$$\nabla^2 \psi = \mu_o\epsilon \frac{\partial^2\psi}{\partial t^2} \tag{2}$$

where the scalar function ψ represents any component of $\overline{E}$ or $\overline{H}$ in Cartesian coordinates.

A fiber is often analyzed using its properties of longitudinal invariance and cylindrical symmetry and assuming a harmonic time dependence. Thus $\overline{E}$ and $\overline{H}$ can be expressed as

$$\begin{bmatrix} \overline{E} \\ \overline{H} \end{bmatrix} = \begin{bmatrix} \overline{E}(\rho,\phi) \\ \overline{H}(\rho,\phi) \end{bmatrix} \exp[i(\omega t - \beta z)] \tag{3}$$

where (ρ, ϕ, z) are the coordinates in the cylindrical system, $\omega = 2\pi f$ is the angular frequency of the lightwave, f is the lightwave frequency, and β is the propagation constant.

Here, for simplicity, it will be assumed that the light wave suffers very little attenuation. This is equivalent to setting ϵ equal to a real number.

The E_z and H_z components are calculated using the wave equation

$$\nabla^2 \begin{bmatrix} E_z \\ H_z \end{bmatrix} = \mu_o\epsilon \frac{\partial^2}{\partial t^2} \begin{bmatrix} E_z \\ H_z \end{bmatrix} \tag{4}$$

which may be written as

$$
\left[\frac{\partial^2}{\partial\rho^2} + \frac{1}{\rho}\frac{\partial}{\partial\rho} + \frac{1}{\rho^2}\frac{\partial^2}{\partial\phi^2} + (\omega^2\mu_o\epsilon - \beta^2) \right] \begin{bmatrix} E_z \\ H_z \end{bmatrix} = 0 \tag{5}
$$

The other components may be calculated using Maxwell's equations giving

$$
\begin{bmatrix} E_\rho \\ \\ H_\phi \\ \\ H_\rho \\ \\ E_\phi \end{bmatrix} = \frac{-i}{\omega^2\mu_o\epsilon - \beta^2} \begin{bmatrix} \dfrac{\omega\mu_o}{\rho} & \beta & 0 & 0 \\ \\ \beta/\rho & \omega\epsilon & 0 & 0 \\ \\ 0 & 0 & \beta & -\dfrac{\omega\epsilon}{\rho} \\ \\ 0 & 0 & -\omega\mu_o & \beta/\rho \end{bmatrix} \begin{bmatrix} \dfrac{\partial H_\phi}{\partial\phi} \\ \\ \dfrac{\partial E_z}{\partial\rho} \\ \\ \dfrac{\partial H_z}{\partial\rho} \\ \\ \dfrac{\partial E_z}{\partial\phi} \end{bmatrix} \tag{6}
$$

Equation 5 is solved using separation of variables. Hence solutions of the form

$$
\begin{bmatrix} E_z \\ H_z \end{bmatrix} = \begin{bmatrix} A \\ B \end{bmatrix} F(\rho) \exp(i\nu\phi) \exp[i(\omega t - \beta z)] \tag{7}
$$

where ν is an integer are sought.

Substituting Equation 7 into Equation 5 leads to Bessel's differential equation for $F(\rho)$:

$$
\left\{ \rho^2\frac{d^2}{d\rho^2} + \rho\frac{d}{d\rho} + \left[(\omega^2\mu_o\epsilon - \beta^2)\rho^2 - \nu^2 \right] \right\} F(\rho) = 0 \tag{8}
$$

A detailed discussion of this equation can be found in references such as Bell's *Special Functions for Scientists and Engineers.*[3]

At this point one needs to consider the optical fiber geometry. A step-index fiber is shown in Figure 1. It is constructed with two concentric cylinders referred to as the core and the cladding. Each region is homogeneous, isotropic, and linear. The refractive indices follow the inequalities given by Equation 9:

$$
n_1 > n_2 \geq n_o \tag{9}
$$

where $n_i^2 = \dfrac{\epsilon_i}{\epsilon_o}$ and ϵ_o is the permittivity of free space. n_o is the refractive index of the material that surrounds the fiber. This is usually taken to be air for which $n_o = 1$. This relation makes propagation by total internal reflection possible as is seen in Section VI.

The geometry of Figure 1 may be simplified to that of Figure 2 by assuming that the cladding radius is much larger than the radius of the core. This is the classical procedure used in the analysis of propagation in fibers.

Using the solutions to Equation 8 appropriate to this simplified fiber geometry,[1] the following equations are obtained:

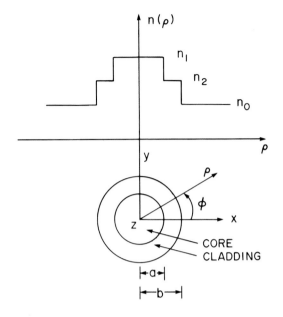

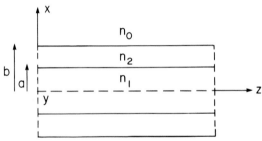

FIGURE 1.   Axial and longitudinal cross-sectional views of a step-index fiber with core of radius a and cladding of radius b. The indices of refraction for the core, cladding, and air are n_1, n_2, and n_0, respectively.

$$\begin{bmatrix} E_z \\ H_z \end{bmatrix} = \begin{bmatrix} A \\ B \end{bmatrix} J_\nu(u\rho) \exp(i\nu\phi) \qquad 0 \leqslant \rho < a \tag{10}$$

where

$$u^2 = \omega^2\mu_0\epsilon_1 - \beta^2 = k_1^2 - \beta^2 \tag{11}$$

and

$$\begin{bmatrix} E_z \\ H_z \end{bmatrix} = \begin{bmatrix} C \\ D \end{bmatrix} K_\nu(w\rho) \exp(i\nu\phi) \qquad a < \rho < \infty \tag{12}$$

where

$$w^2 = \beta^2 - \omega^2\mu_0\epsilon_2 = \beta^2 - k_2^2 \tag{13}$$

The functions J_ν and K_ν are Bessel functions and modified Hankel functions, respectively. The constants A, B, C, and D in Equations 10 and 12 are determined using the boundary

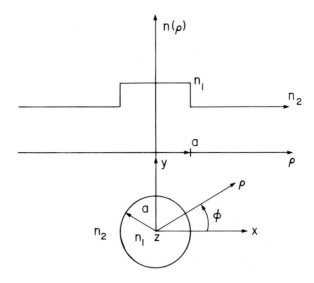

FIGURE 2.   Fiber of Figure 1 with cladding radius $b \to \infty$.

conditions for the tangential field components at the core-cladding interface. The parameters k_1 and k_2 are propagation constants for plane waves propagating in lossless media with permittivities ϵ_1 and ϵ_2, respectively, and β is the propagation constant for the fields.

The parameters u and w are nonnegative. Thus, the following relationship may be stated:

$$k_o^2 n_2^2 \leq \beta^2 \leq k_o^2 n_1^2 \tag{14}$$

where n_1, n_2 are the refractive indices of the core and the cladding, respectively, and $k_o = \omega \sqrt{\mu_o \epsilon_o} = 2\pi/\lambda$ is the wave number of free space. λ is the free space wavelength.

From Equations 10 and 12, all the other field components may be calculated using Equation 6 in both the core and the cladding. Applying the boundary conditions on tangential E and H to these results[1] yields a system of homogeneous equations in A, B, C, and D. Nonzero solutions of these systems of equations occur only when the determinant of this set of equations is equal to zero (Equation 15). This condition is

$$\begin{vmatrix} J_\nu(ua) & 0 & -K_\nu(wa) & 0 \\[2mm] 0 & J_\nu(ua) & 0 & -K_\nu(wa) \\[2mm] \dfrac{i\nu\beta}{u^2 a} J_\nu(ua) & -\dfrac{\omega\mu}{u} J_\nu'(ua) & \dfrac{i\nu\beta}{w^2 a} K_\nu(wa) & -\dfrac{\omega\mu}{w} K_\nu'(wa) \\[2mm] \dfrac{\omega\epsilon_1}{u} J_\nu'(ua) & \dfrac{i\nu\beta}{u^2 a} J_\nu(ua) & -\dfrac{\omega^2\epsilon_2}{w} K_\nu'(wa) & \dfrac{i\nu\beta}{w^2 a} K_\nu'(wa) \end{vmatrix} = 0 \tag{15}$$

This equation may be solved for the propagation constant β. It is usually rewritten in the following form:

$$(\epsilon f + g)(f + g) - \nu^2 \left(\frac{\epsilon}{U^2} + \frac{1}{W^2} \right) \left(\frac{1}{U^2} + \frac{1}{W^2} \right) = 0 \tag{16}$$

where

$$f = \frac{J_\nu'(U)}{UJ(U)}$$

$$g = \frac{K_\nu'(W)}{WK_\nu(W)} \tag{17}$$

$$U^2 = (ua)^2 = (k_1a)^2 - (\beta a)^2 \tag{18}$$

$$W^2 = (wa)^2 = (\beta a)^2 - (k_2a)^2 \tag{19}$$

and

$$\epsilon = \frac{\epsilon_1}{\epsilon_2} = \frac{n_1^2}{n_2^2}$$

$$k_i = \omega^2\mu\epsilon_i = k_O^2 n_i$$

$$i = 1,2 \tag{20}$$

$$J_\nu'(U) = \frac{1}{2}[J_{\nu-1}(U) - J_{\nu+1}(U)] \tag{21}$$

$$K_\nu'(W) = -\frac{1}{2}[K_{\nu-1}(W) + K_{\nu+1}(W)] \tag{22}$$

The above Equations (15) or (16) are called the exact characteristic equations. Each value of β that satisfies it represents a different guided mode.[1]

Depending on the value of ν, these modes have various properties. If $\nu = 0$, for example, TE and TM modes may exist in the fiber. If $\nu \geq 1$, a set of hybrid modes called HE and EH modes will propagate in the fiber.

An important parameter for each mode is its cutoff frequency. A mode is said to be cutoff when its field does not decay outside of the core. In a geometrical sense, this means that the fields associated with this mode are not totally internally reflected at the core-cladding interface. Rather, they are partially transmitted, and there is a continuous radiation loss.

It can be shown[1] that $W \rightarrow O$ is the mode cutoff condition. The values of U in the characteristic equation associated with the parameter $W \rightarrow O$ are called the cutoff values of U or U_c. The cutoff frequency, f_c, is given by

$$f_c = \frac{cU_c}{2\pi a[n_1^2 - n_2^2]^{1/2}} \tag{23}$$

where

$$c = \frac{1}{\sqrt{\mu_o\epsilon_o}} \tag{24}$$

A summary of the modes in a step-index fiber and their cutoff equations is given in Table 1. Here the integer μ is used to enumerate the various modes associated with a given value of ν. Note that μ used here is not the permeability defined earlier.

It is important to note that there is one mode that possesses the cutoff value $U_c = 0$ and, thus, a cutoff frequency equal to zero. This is because $J_n(0) = 0$, $n \neq 0$. This mode is the

Table 1
MODES IN A STEP-INDEX OPTICAL FIBER

ν	Cutoff equations	Modes
1	$J_1 (U_c) = 0$	$HE_{1\mu}$
≥ 1	$J_\nu(U_c) = 0; U_c \neq 0$	$EH_{\nu\mu}$
>1	$(\epsilon + 1) J_{\nu - 1} (U_c) - U_c J_\nu (U_c)/(\nu - 1) = 0$	$HE_{\nu\mu}$
0	$J_0 (U_c) = 0$	$TE_{0\mu}, TM_{0\mu}$

HE_{11} mode that is called the fundamental or dominant mode. Optical fibers that permit only propagation of this mode are called single-mode or monomode fibers. Single-mode operation occurs in the frequency range:

$$0 < f < \frac{2.4048\ c}{2\pi a\ [n_1^2 - n_2^2]^{1/2}} \tag{25}$$

where the value 2.4048 is the cutoff value of the next higher modes (TE_{01} and TM_{02}).

Fibers that are operated with any frequency at or higher than the limit given in Equation 25 are called multimode optical fibers.

To solve the characteristic equation (Equation 15), numerical methods can be used. In doing so, it is convenient to define the parameter V, a normalized frequency, as:

$$V^2 = U^2 + W^2 \tag{26}$$

This parameter may be written in the form:

$$V = \frac{2\pi a}{\lambda}\ [n_1^2 - n_2^2]^{1/2} \tag{27}$$

For a given set of refractive indices, curves of β/k_0 can be plotted for representative modes by solving the characteristic equation (Equation 15). A representative result is shown in Figure 3. As the radius of the fiber core is increased, a greater number of modes are guided by the structure. Changing the value of n_1 and n_2 will alter only the vertical scale and the cutoff values of the $HE_{\nu\mu}$ modes ($\nu > 1$).

Each mode may have a different field configuration and may present a different propagation constant to incident light energy. This gives rise to bandwidth limitations as discussed later in the chapter.

It should be noted that in almost all cases, the difference between n_1 and n_2 is only a few percent or less. When this is the case, an approximation to the characteristic equation called the "weakly guiding approximation" can be made.[2] When this approximation is made, the modes are labeled differently and called LP (linearly polarized) modes.

III. MATERIAL DISPERSION IN STEP-INDEX FIBERS

As pulses of light propagate in an optical fiber, they become distorted. The distortion is due in part to the fact that light pulses contain a spectrum of wavelengths. Both the finite line width of the light source and spectral broadening due to modulation are responsible for the spectrum of the light pulses, although the former usually dominates.

For distortionless propagation of these pulses in monomode fibers, the propagation constant β of the HE_{11} mode as defined in Section II must be a linear function of frequency.*

* For rays the propagation constant is $k_0 n$ as defined in Section VI.

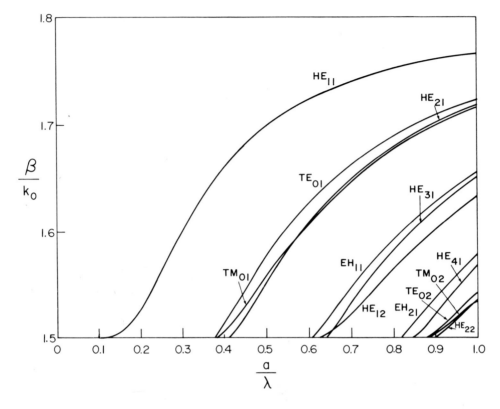

FIGURE 3.   Shows the function (β/k_o) vs. (a/λ) for a collection of lower modes in step-index fibers. Note that $n_1 = 1.8$ and $n_2 = 1.5$.

This can be shown in the following way. Let $g_i(t)$ be a pulse input signal to the fiber. $G_i(\omega)$ is its Fourier transform.*   According to Equation 7 the pulse, after propagating a distance z in the fiber, is

$$g_o(t) = \frac{1}{2\pi} \int_{-\infty}^{\infty} \hat{G}_i(\omega) \exp[i(\omega t - \beta(\omega)z)]d\omega \qquad (28)$$

$\beta(\omega)$ may be expanded in a Taylor series around the center frequency ω_c of the pulse

$$\beta(\omega) = \beta(\omega_c) + \frac{\partial\beta}{\partial\omega}\bigg|_{\omega=\omega_c} (\omega - \omega_c) + \ldots \qquad (29)$$

Now, *if $\beta(\omega)$ is a linear function of frequency, then the Taylor series consists of only two terms and*

$$g_o(t) = \frac{1}{2\pi} \int_{-\infty}^{\infty} \hat{G}_i(\omega) \exp[i\omega(t - \frac{\partial\beta}{\partial\omega} z)] \, d\omega$$

$$= g_i(t - \frac{\partial\beta}{\partial\omega} z) \qquad (30)$$

---

* Due to the nature of optical sources, $g_i(t)$ is not deterministic as assumed here. In a more correct development, $g_i(t)$ is treated as a sample function and statistical averaging is done to obtain the light intensity. Nevertheless, deterministic theory can be used to illustrate the distortion phenomenon.

This is an undistorted replica of the input signal. (Note that the constant term has been dropped because it does not affect the pulse shape.)

Generally, however, β is a nonlinear function of frequency and distortion is introduced into the pulse by the higher order terms of the Taylor series.

For multimode fibers, distortionless propagation occurs only if the propagation constant for each mode is identical *and* is linear with frequency. Distortion due to the presence of multiple modes is usually described as intermodal. It is discussed in Sections VII and IX and will not be considered further here.

Distortion in single-mode fibers (or due to any particular mode in a multimode fiber) is usually called intramodal. The nonlinearities in the propagation constant that are responsible for this type of distortion can be traced to two sources. The first is that the refractive indices of the core and the cladding are functions of frequency. This phenomenon leads to a nonlinear variation of β with frequency. This source of distortion is called material dispersion and is the subject of this section. The second intramodal effect occurs even if the core and cladding materials have frequency independent refractive indices. If this is so, β will still be a nonlinear function of frequency due to waveguiding effects. This effect is called waveguide dispersion and will be combined with material dispersion in Section V for the single-mode case.

It should be noted that, in the 0.8 to 0.9 μm wavelength range used by many optical fiber systems, material dispersion dominates waveguide dispersion. Since this is the case, and since the analysis of waveguide dispersion is complicated, it is useful to examine material dispersion separately. This can be done by studying the propagation of plane waves in a homogeneous medium. In this case, there are no waveguiding effects because the wave is not guided by boundaries. This approach will be used here since it is simple and can be used to approximately predict total intramodal dispersion in the important 0.8 to 0.9 μm wavelength range. As mentioned above, the wavelength range for which waveguide dispersion is important will be discussed in Section V.

Monochromatic plane electromagnetic waves characterized by the wave equation propagate through a homogeneous medium with permittivity ϵ and permeability μ with the phase velocity v_p defined by:

$$v_\rho = \frac{1}{\sqrt{\mu\epsilon}} = \frac{\omega}{k_o n} \tag{31}$$

where: $\omega = k_o c$ is the radian or angular frequency, in rad/sec, of the source; c is the phase velocity of a wave in free space (speed of light); and n is the refractive index of the medium, given by

$$n = \frac{c}{v_\rho} \tag{32}$$

Normally, optical systems involve nonmagnetic media. Thus n can be written as:

$$n = \sqrt{\epsilon_r}$$

where

$$\epsilon_r = \epsilon/\epsilon_o$$

In dispersive media such as optical fibers, n (or ϵ_r) depends on the frequency or wavelength of the incident wave. In such media, the refractive index is not constant with wavelength, and the wavelength dependence is usually nonlinear. In Section V, this dependence will be characterized using the three-term Sellmeier equation.[4]

Fiber optic communication systems based on intensity modulation are subject to dispersion as it affects the envelope of a propagating pulse.[5] The propagation velocity of this envelope is known as the group velocity, v_g, given by:

$$v_g = 1 / \frac{\partial \beta}{\partial \omega} \tag{33}$$

This definition comes from Equation 30. The group and phase velocities are only equivalent in nondispersive media (β linear with ω).

Using the relation $\beta = \omega \, n/c = 2\pi n/\lambda$, v_g can be rewritten in terms of derivatives with respect to wavelength

$$v_g = \frac{c}{n - \lambda \dfrac{dn}{d\lambda}} = \frac{c}{N} \tag{34}$$

where

$$N = n - \lambda \frac{dn}{d\lambda} \tag{35}$$

In the above equation N is called the group index of the propagation medium.

Gloge[5] defines the group delay, τ, as the inverse of the group velocity. Thus, using Equations 33 and 34,

$$\tau = \frac{1}{v_g} = \frac{N}{c} = \frac{d\beta}{d\omega} = \frac{1}{c} \frac{d\beta}{dk_o} \tag{36}$$

The parameter τ indicates physically the pulse envelope delay per unit of fiber length. τ may account for material and waveguide dispersion effects if β is obtained from the characteristic equation, (Equations 16 to 22). In this case, N represents the total group index N_T for the fiber as will be seen in Section V.

The transmission time, T, for a given optical pulse transmitted over a length L of optical fiber is

$$T(\lambda) = L\tau(\lambda) = \frac{LN(\lambda)}{c} \tag{37}$$

where again λ is the light wavelength in free space.

To this point the pulse spectrum has been assumed narrow enough that Equation 29 can be truncated at two terms. This is not generally true for optical sources. One method for circumventing this problem is to assume that the light source (of spectral width $\Delta\lambda$) consists of a set of ideal monochromatic sources covering the wavelength range λ to $\lambda + \Delta\lambda$. Each monochromatic carrier is modulated by the same pulse signal. Because the spectrum of the pulse signal is narrow compared to $\Delta\lambda$, each modulated carrier is narrow band enough that a two-term approximation to Equation 29 is adequate. Each wavelength component of the source's output will propagate along the fiber with a different group velocity because v_g is a function of wavelength. Thus, the input pulse to the fiber will arrive at the optical detector distorted or broadened.

The difference in transit times for the wavelength components at each end of the spectrum will be

$$\Delta T(\lambda, \lambda + \Delta\lambda) = T(\lambda + \Delta\lambda) - T(\lambda)$$

$$= \frac{L}{c} [N(\lambda + \Delta\lambda) - N(\lambda)] \tag{38}$$

Expanding $N(\lambda + \Delta\lambda)$ in a Taylor series gives

$$\Delta T(\lambda, \lambda + \Delta\lambda) = L \sum_{m=1}^{\infty} D_m(\Delta\lambda)^m \tag{39}$$

where

$$D_m = \frac{1}{m!c} \frac{d^m N}{d\lambda^m} \tag{40}$$

The D_m coefficients are called dispersion coefficients.[6] The first order dispersion coefficient is:

$$D_1 = \frac{1}{c} \frac{dN}{d\lambda} \tag{41}$$

while the second order dispersion coefficient is:

$$D_2 = \frac{1}{2c} \frac{d^2 N}{d\lambda^2} \tag{42}$$

For optical sources with narrow spectra, such as semiconductor diode lasers, the ratio $\Delta\lambda/\lambda$ is much less than one, and the higher order terms ($m > 2$) in Equation 39 can be ignored.[7]

The above problem can be attacked from another point of view. Consider that the source has a spectral width $\Delta\lambda$ centered at the wavelength λ chosen so that the first order dispersion coefficient is zero:

$$D_1 = \frac{1}{c} \frac{dN}{d\lambda} \bigg|_{\lambda=\hat{\lambda}} = 0 \tag{43}$$

By writing the required Taylor series for ΔT under these new conditions:

$$\Delta T(\hat{\lambda} + \Delta\lambda/2, \hat{\lambda} - \Delta\lambda/2) = T(\hat{\lambda} + \Delta\lambda/2) - T(\hat{\lambda} - \Delta\lambda/2)$$

$$= \frac{L}{c} [N(\hat{\lambda} + \Delta\lambda/2) - N(\hat{\lambda} - \Delta\lambda/2)] \tag{44}$$

Here

$$N(\hat{\lambda} \pm \Delta\lambda/2) = N(\hat{\lambda}) \pm N'(\hat{\lambda})\Delta\lambda/2$$

$$+ \frac{1}{2} N''(\hat{\lambda}) (\Delta\hat{\lambda}/2)^2 + \ldots \tag{45}$$

where

$$N'(\hat{\lambda}) = \frac{dN}{d\lambda}\bigg|_{\lambda = \hat{\lambda}} \tag{46}$$

Thus

$$\Delta T \simeq \frac{L}{c} [N'(\hat{\lambda})\Delta\lambda] \tag{47}$$

With the assumptions of: (1) that $\Delta\lambda/\lambda << 1$, and (2) that a $\lambda = \hat{\lambda}$ exists such that $N'(\lambda) = 0$,

$$\Delta T = 0 \tag{48}$$

Thus at the wavelength $\hat{\lambda}$, optical frequency components from opposite sides of the spectrum have the same group velocity and will arrive at the receiver at the same time, given the above assumptions. Later, in Section V, it will be shown that the assumption $D_1 = 0$ is at the heart of classical theory for monomode fiber system optimization.

At the wavelength $\hat{\lambda}$, the worst-case ΔT, ΔT_{max}, occurs when the effects of spectral components at $\lambda = \hat{\lambda}$ and $\lambda = \hat{\lambda} + \Delta\lambda/2$ are compared while $D_1 = 0$. Here

$$\begin{aligned}
\Delta T_{max} &= \Delta T(\hat{\lambda}, \hat{\lambda} + \Delta\lambda/2) \\
&= \frac{L}{c} [N(\hat{\lambda} + \Delta\lambda/2) - N(\hat{\lambda})] \\
&= \frac{L}{2c} (\Delta\lambda/2)^2 \frac{d^2N}{d\lambda^2}\bigg|_{\lambda = \hat{\lambda}} \\
&= L\hat{D}_2(\Delta\lambda/2)^2 \tag{49}
\end{aligned}$$

where

$$\hat{D}_2 = D_2/4 = \frac{1}{8c} \frac{d^2N}{d\lambda^2}\bigg|_{\lambda = \hat{\lambda}} \tag{50}$$

The above factors and coefficients are well-known in the literature. D_1 is often called simply the dispersion. The wavelength $\hat{\lambda}$ is called the zero dispersion wavelength, and the coefficient D_2 is called the residual dispersion.

Normally D_1 is stated in psec/(km − nm) and D_2 is given in psec/(km − nm²). In Section V a design philosophy for monomode fibers that makes use of D_1 will be discussed.

For the materials usually used to construct optical fibers, a zero dispersion wavelength can be found in the vicinity of 1.3 μm. Near this wavelength the value of the material dispersion is controlled by D_2. Further, waveguide dispersion is an important component of intramodal dispersion, as will be discussed in Section V. Because a zero in material dispersion occurs near 1.3 μm, this wavelength is of great interest for designers of high data rate optical fiber systems.

IV. ATTENUATION IN FIBERS

The widespread use of optical fibers is due mainly to progress that has been made in the last few years in bringing fiber attenuation from hundreds of decibels per kilometer to a decibel or less per kilometer. It is well-known that this low level of attenuation is available if one selects optical sources in the 1.0 to 1.6 μm range. Minimum values are generally achievable at specific wavelengths depending on the material used, the manufacturing process, the cladding thickness, and the number of modes excited in the fiber. Generally, windows of minimum attenuation are encountered at about 1.20, 1.30, and 1.55 μm. High attenuation regions usually exist at about 1.25 and 1.38 μm.

For monomode fibers, it is of interest to design systems such that the wavelength for minimum attenuation and the zero dispersion wavelength are the same. It has been suggested[8] that an optical source with λ = 1.55 μm can be appropriately matched to a low-loss fiber to accomplish this.

The losses in the fiber itself can be ascribed to several sources.[2] The glass used in the fiber introduces losses to the system due to the intrinsic material absorption and due to Rayleigh scattering of the light wave. Additional losses can be introduced by the presence of metal ions, but modern techniques of production have virtually eliminated this as a cause of significant fiber attenuation. Another impurity ion, the OH radical, is well-known as the cause of the attenuation maxima at 1.25 and 1.38 μm as well as the high attenuation seen between 0.9 and 1.0 μm.

Modification of fiber material characteristics due to exposure to temperature extremes or intense ionizing radiation can dramatically increase attenuation, although in some cases this may be reversible.[2]

Reduction of attenuation in any optical fiber is limited by the Rayleigh scattering losses due to the minute index of refraction fluctuations present at all times in optical fibers. This phenomenon probably represents the lower boundary of fiber attenuation loss for shorter wavelengths (λ < 1.6 μm). The Rayleigh scattering loss is inversely proportional to the factor λ.[4] This dependence on λ is responsible for the fact that fibers operated at longer wavelengths generally have smaller losses. Typical losses for good quality fibers at 0.85 μm are 4 to 5dB/km while losses for good quality fibers at 1.3 μm are less than 1 dB/km. This lower loss is one of the advantages of longer wavelength systems. For λ > 1.6 μm, ultraviolet absorption increases dramatically.

V. MONOMODE FIBER DESIGN

Designing optimal monomode systems involves basically the expression for the first order dispersion coefficient (Equation 41). The difference between the analysis here and that given in Section III is that both fiber material dispersion and the wavelength dependence of the group delay of the single propagating mode (waveguide dispersion) will be considered simultaneously. If the system can be designed so that $\hat{\lambda}$ (the zero dispersion wavelength) occurs when attenuation is minimal in the fiber, so much the better.*

The first effect mentioned above depends only on the materials used in the fiber. Customarily, one assumes that the refractive indices of both core and cladding, n_1 and n_2, respectively, follow the three-term Sellmeier Equation:[4]

$$n_j^2 = 1 + \sum_{i=1}^{3} \frac{A_i}{\lambda^2 - l_i^2} \qquad j = 1, 2 \tag{51}$$

where A_i are constants related to the number of particles in the material that can oscillate at wavelengths l_i.

The second effect, called waveguide dispersion, is a consequence only of the waveguiding properties of the optical fiber. This dispersion depends on the core radius, the propagation constant of the selected dominant mode (HE_{11}) and some of its derivatives.

The combination of the above two dispersive effects gives the total dispersion (or, simply,

---

* Note that $\hat{\lambda}$ is now different than in Section III because waveguide and material dispersion are treated together.

dispersion) of pulses bounded in a single-mode fiber. Since this dispersion limits the useful bandwidth of optical communication systems, one should use a wavelength that minimizes this effect.

The wavelength for minimum dispersion can be computed using asymptotic formulas. This approach has been used because, in most cases, the relative difference of refractive indices of the core and cladding given by*

$$\Delta = (n_1 - n_2)/n_2 \tag{52}$$

is small.

The condition $\Delta \ll 1$ allows a substantial simplification of the exact characteristic equation (Equation 15) that is the starting point for the dispersion analysis of the selected optical fiber. Two asymptotic procedures have been employed in the past.

1.   One involves working with the values of the dominant mode propagation constant obtained directly from the solution of the approximate characteristic equation and using, first, analytical expressions and, then, numerical methods to obtain the values of some of the derivatives of the dominant mode propagation constant.
2.   The other involves working with approximations for the parameters of the approximate characteristic equation (these are analytically simple formulas) and using these expressions to obtain the values of the HE_{11} mode propagation constant and some of its derivatives.

However, it is possible to obtain $\hat{\lambda}$ directly using analytical and computational procedures. Total dispersion, as discussed above, is given by:[9]

$$D_1 = \frac{1}{c} \frac{dN_T}{d\lambda} \tag{53}$$

where c is the speed of light in free space and N_T is the total group index. N_T is given below in Equation 54. It is specified in terms of a series of parameters. In particular, b is the normalized propagation constant for the HE_{11} mode, V is a normalized frequency, N_i (i = 1, 2) are the group indices of the core and cladding, respectively, and n_e is the effective phase index, i.e., the phase index "seen" by the HE_{11} mode propagating in the optical fiber under consideration. Thus, it is evident that N_T includes both material and waveguide dispersion:**

$$N_T = \partial\beta/\partial k_o$$

$$= \frac{1}{n_e} \left[n_2 N_2 + \left(\frac{V}{2} \frac{db}{dV} + b \right) \theta \right] \tag{54}$$

where

$$\theta = n_1 N_1 - n_2 N_2 \tag{55}$$

$$N_i = n_i - \lambda \frac{dn_i}{d\lambda}$$

$$i = 1, 2 \tag{56}$$

$$n_e = [n_2^2 + (n_1^2 - n_2^2)b]^{1/2} \tag{57}$$

---

*   The definition of $\triangle$ used in this section is slightly different from that used in later sections.   In most practical cases, the two definitions give almost identical results.

**   If a mode is far from cutoff, then U $\ll$ V and, from Equation 58, b $\simeq$ 1. Using this fact, Equation 54 reduces to Equation 35, which was derived for plane wave propagation in a homogeneous medium. Thus, if a mode is far from cutoff, the analysis of Section III can be used and waveguide dispersion ignored.

The normalized propagation constant is given by

$$b = W^2/V^2 = 1 - U^2/V^2 \tag{58}$$

The parameters U (or W) come from the solution of the exact characteristic equation written more conveniently as[1]

$$(J^+ + K^+)(\epsilon J^- - K^-) + (J^- - K^-)(\epsilon J^+ + K^+) = 0 \tag{59}$$

where

$$J^+ = \frac{J_{\nu+1}(U)}{U J_\nu(U)} \qquad J^- = \frac{J_{\nu-1}(U)}{U J_\nu(U)} \tag{60}$$

$$K^+ = \frac{K_{\nu+1}(W)}{W K_\nu(W)} \qquad K^- = \frac{K_{\nu-1}(W)}{W K_\nu(W)} \tag{61}$$

using ν equal to unity. In Equations 60 and 61, J and K are Bessel functions and modified Hankel functions, respectively. The electrical permittivity in Equation 59 is given by

$$\epsilon = n_1^2/n_2^2 \tag{62}$$

Taking into account the wavelength dependence of the parameters involved in the total group index, Equation 53 becomes[9]

$$D_1 = \frac{1}{c}\left(\frac{A_1 A_2 - A_3 A_4}{A_5}\right) \tag{63}$$

where

$$A_1 = n_e \tag{64}$$

$$A_2 = \phi_2 + (\phi_1 - \phi_2)\left(\frac{V}{2}\frac{db}{dV} + b\right)$$

$$\quad - \frac{V\,\theta^2}{2\lambda(n_1^2 - n_2^2)}\left(V\frac{d^2b}{dV^2} + 3\frac{db}{dV}\right) \tag{65}$$

$$A_3 = n_2 N_2 + \left(\frac{V}{2}\frac{db}{dV} + b\right)\theta \tag{66}$$

$$A_4 = \frac{1}{n_e}\left[(1 - b)n_2 n_2' - \frac{V\theta}{2\lambda}\frac{db}{dV} + n_1 n_1' b\right] \tag{67}$$

$$A_5 = n_e^2 \tag{68}$$

and

$$\phi_j = N_j n_j' - \lambda n_j n_j'' \tag{69}$$

$$n_j' = -\frac{1}{n_j}\sum_{i=1}^{3}\frac{A_i\, l_i^2\, \lambda}{(\lambda^2 - l_i^2)^2} \tag{70}$$

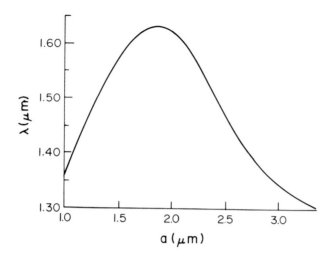

FIGURE 4.   Variation of the wavelength for total minimum dispersion
$\hat{\lambda}$ as a function of the core radius a. This calculation is based on Sample
2, Table 1 of Reference 9 using a value of Δ = 0.6%.

$$n_j^{''} = \frac{1}{n_j} \left[- (n_j^{'})^2 + \sum_{i=1}^{3} \frac{A_i \, l_i^2 (3\lambda^2 + l_i^2)}{(\lambda^2 - l_i^2)} \right] \tag{71}$$

The prime on n_j indicates differentiation with respect to wavelength. The subscript j can
be either 1 or 2.

To operate the fiber at the maximum transmission rate, we must select the wavelength
that corresponds to minimum total dispersion. Thus we select λ such that

$$D_1 \Big|_{\lambda = \hat{\lambda}} = 0 \tag{72}$$

Computational procedures for solving Equation 72 are described by Pires et al.[9] A basic
result from this procedure is shown in Figure 4. Here one observes the variation of $\hat{\lambda}$ with
core radius, using a value of Δ = 0.006. From the figure it is possible to design a monomode
optical fiber with a step-index profile to give minimum dispersion for the wavelength of the
available optical source by selecting the appropriate core radius.

Figure 5 shows the variation of material (D_m), waveguide (D_g), and total (D_1) dispersion
for a fiber with Δ = 0.0215 and 2a = 3.63 μm according to the approach outlined here
or in Reference 9 (labeled EXACT), and according to Chang's approximate method.[8,10] Note
that the zero dispersion values occur at different frequencies, and that a proper design is
based on D_1 = 0.

Having considered the monomode case, the interesting problems encountered in the multimode fiber will now be discussed.

VI. INTRODUCTION TO RAY THEORY FOR MULTIMODE FIBERS

In Section V it has been assumed that there is only one propagating mode in the fiber.
While monomode fibers are attractive for long distance communication systems, they are
not always used for the following reason. The coupling of energy between incoherent light
sources (e.g., LEDs) and monomode fibers is very poor because the core of a monomode
fiber has a very small diameter.[11] Monomode fibers, however, have a much larger diameter

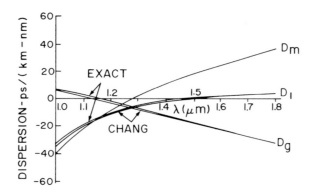

FIGURE 5.   Dispersion curves for a fiber constructed using Sample 2 material.[9] D_m is the material dispersion, D_g is the waveguide dispersion, and D_l is the total dispersion. These calculations were made using a core diameter of $2a = 3.63$ μm and a relative difference $\Delta = 2.15\%$.

core and can be coupled to LEDs without excessive loss. Since LEDs are inexpensive, highly reliable, and adequate for many low data rate communication applications, it is desirable to use a fiber with which they are compatible.

For these reasons, multimode fibers are very important. As the name implies, multimode fibers are fibers in which many modes can propagate. The fundamental difference in the construction of multimode and monomode fibers is that the multimode fiber core has a diameter that is at least ten times larger than that of the monomode fiber core. Since the number of propagating modes is proportional to the square of the core diameter, it is evident that more than a hundred modes may propagate in typical multimode fibers.[12]

The energy carried by a pulse propagating in a multimode fiber is distributed among the modes. Since, in general, each mode has a distinct group velocity, the portion of the energy carried by each mode will have a distinct arrival time at the detector end of the fiber. It is evident that this distribution of arrival times leads to pulse distortion and hence to intersymbol interference, which is in addition to the effects of material and waveguide dispersion, discussed in Sections III and V. This additional phenomenon is called "intermodal" dispersion.

While it is possible and common to describe the propagation of light in a fiber by superimposing the contribution of each propagating mode, it is difficult when the number of modes is large. Further, the mathematical complexity inherent in determining the properties of modes can obscure the essential physics of the propagation process.

An alternative to the modal description of the propagation process is the ray description. In this description, the propagation of light between two points is described by assuming that the light travels along paths that are governed by the laws of geometric optics. These paths are called ray paths. Along each ray path, the field intensity in a low-loss fiber varies as

$$\frac{\exp(-jk_ons)}{s} \tag{73}$$

where $k_o = 2\pi/\lambda$ is the propagation constant of free space, λ is the free space wavelength of the source, and s is the total distance along the ray path measured from the source. The parameter n is the refractive index of the medium and may be complex to account for losses. To describe propagation of light, all ray paths that connect the source and detector must be identified. In general, this will be a tedious process. However, to predict the amount of pulse broadening that results from multiple ray path propagation, only two ray paths must

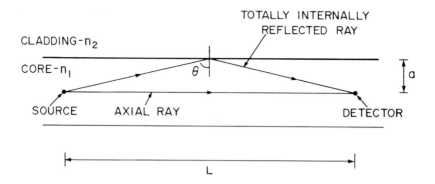

FIGURE 6.   Example ray paths between source and detector in a step-index multimode fiber.

be identified. The details of this calculation will be presented later. Another attractive characteristic of ray theory is that it leads to an intuitive understanding of propagation in different types of optical fibers.

Several things should be noted about the mode and ray theories. First and most important is that the two theories can be shown to yield equivalent results provided that ray theory is suitably generalized.[13] Ray theory is more easily used, however, in its simple form as outlined above. While the simple ray theory is strictly valid only for a wavelength of zero, it gives good results in most cases when the wavelength is small compared to the dimensions of the fiber. For multimode fibers, this is the case since a typical core diameter is approximately 50 times the wavelength of light. A discussion of some of the differences between the simple ray theory and mode theory is given by Miller and Chynoweth.[14]

VII. RAY DISPERSION IN MULTIMODE STEP-INDEX FIBERS

The simple ray theory will be used here to describe propagation in step-index fibers. Such a fiber is shown in Figure 1. The fiber is separated into two homogeneous regions: the core and the cladding. The refractive indices of the core and cladding are, respectively, n_1 and n_2 where it is assumed that $n_1 > n_2$. The diameter of the core is 2a. It will be assumed that the cladding-air interface can be ignored as in Figure 2.

The first simplification that will be made is that only meridional rays will be considered. Meridional rays are rays that intersect the axis of the fiber as shown in Figure 6. Nonmeridional rays are called skew rays and have ray paths that encircle the fiber axis. Neglecting skew rays will not cause a qualitative change in the conclusions made here about fiber characteristics. The effect of including skew rays is discussed in more detail in several references.[14,15]

According to the laws of geometric optics, a ray propagating in a homogeneous medium travels in a straight line. At an interface between two media, however, the wave nature of light must be considered since the refractive index varies significantly within the span of a wavelength. Here the light may be treated as a plane electromagnetic wave that will be reflected from and/or transmitted through the interface. The laws that govern this reflection are Snell's laws and the Fresnel reflection coefficients that have been discussed in Chapter 1. Of most interest here is the total internal reflection phenomenon. As mentioned in Chapter 1, total internal reflection occurs for rays that intersect the interface at an angle

$$\theta > \theta_c = \sin^{-1}(n_2/n_1) \tag{74}$$

where θ_c is the critical angle. When this condition is satisfied, no energy is transmitted through the interface. All of the energy carried by the incident ray is reflected back into the core.* The result is the zig-zag ray path shown in Figure 6. Rays that satisfy this condition are called "bound" rays. It should be noted that, as shown in Chapter 1, the boundary conditions require a field to exist in the cladding that decays exponentially away from the interface. Thus the cladding must be chosen to be large enough so that the field will be negligible at its outside boundary.

The number of ray paths that can be found depends upon the relative locations of the source and detector, the diameter of the core, and the ratio n_1/n_2.** It is not always necessary, however, to identify all of the ray paths. The most important characteristic of propagation in multimode fibers can be approximately calculated by referring only to the two limiting rays (i.e., the ray that travels along the fiber axis and the ray that intersects the core-cladding interface at the critical angle). The optical path length in a homogeneous medium is defined as the product of the refractive index n and the ray path length. The optical path length in the step-index fiber is derived using the fact that the ray path lengths for the two limiting rays in Figure 6 are L and n_1L/n_2, respectively. Thus, rays traveling a distance L along the fiber axis accumulate an optical path length n_1L while rays that travel along a zig-zag path at the critical angle accumulate an optical path length $n_1^2 L/n_2$. The difference between these two optical path lengths is the maximum difference in optical path length for bound rays.

The ray transit time (optical path length divided by the speed of light) is defined as the time it takes a wavefront to travel along a given ray path through a fiber of length L. Here, the variation of refractive index with wavelength that leads to material dispersion as discussed in Section III has been ignored. When this is done, the phase velocity and the group velocity of the ray are identical. Thus, for the step-index fiber, an axial ray has a transit time

$$t_{min} = n_1L/c \tag{75}$$

This is the minimum time required for a bound ray to travel a distance L in the fiber. A ray that travels at the critical angle has a transit time

$$t_{max} = n_1^2 L/(n_2c) \tag{76}$$

This is the maximum transit time for bound rays. The difference between these two times is a measure of the pulse broadening due to ray dispersion. The pulse broadening is defined here as the amount of time by which the pulse length is increased. For a step-index fiber, then, the pulse broadening can be written†

$$T_d(L) = n_1^2 L\Delta/(n_2c) \tag{77}$$

where $\Delta = (n_1 - n_2)/n_1$. It is evident from this equation that pulse broadening can be decreased by decreasing Δ. This, in fact, is one of the reasons that the core-cladding refractive index difference is chosen to be small.

---

* In terms of modal theory, the light that is not totally reflected can be described as being contained in "radiation" modes or approximately as a finite number of "leaky" modes.[16] Although these modes will not be discussed further, they may be important in at least two cases. First, Snyder has shown that, due to their small attenuation, "tunneling leaky modes" may play an important role in the description of propagation in round fibers.[17] Second, it is evident that at bends in the fiber, some rays which were totally internally reflected in the straight fiber now approach the interface at less than the critical angle and are at least partially transmitted (i.e., radiated). This phenomenon is responsible for radiation at bends in the fiber.[14]

** If the source detector distance is large compared with the core diameter, then essentially all angles between $\pi/2$ and the critical angle are possible.

† Equation 77 is not the usual measure of pulse broadening. "RMS pulse broadening" which accounts for all rays (or modes) is the more widely used.[12] For step-index fibers, RMS pulse broadening is smaller than that given in Equation 77 by a factor of $2\sqrt{3}$.

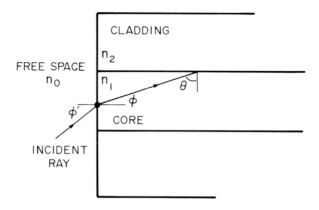

FIGURE 7. Source-fiber coupling in a multimode optical fiber.

A typical RMS pulse broadening for a step index-fiber is 20 nsec/km. This can be compared to a typical RMS pulse broadening of 0.2 nsec/km due to material dispersion for a laser source at a wavelength of 0.82 μm or 2 nsec/km for an LED source at the same wavelength. It is clear that at this wavelength, ray dispersion usually dominates material dispersion in step-index fibers.

VIII. SOURCE-FIBER COUPLING AND NUMERICAL APERTURE

As mentioned above, it is possible to decrease pulse broadening by decreasing Δ. However, as Δ is decreased, the amount of power that can be coupled from a source into the bound rays of a fiber is decreased. This effect decreases the power available to the detector and, as a result, adversely affects the system error rate. Thus there is a lower limit for the value of Δ that can be used.

One measure of the coupling efficiency between sources and fibers is the fiber numerical aperture (NA). The NA of a fiber can be calculated in the following way. Assume that there is a point source of light at the end of and on the axis of a fiber. Light rays are radiated from the point source equally in all directions. Only light incident on the end of the fiber in certain directions, however, is coupled into the bound rays of the fiber. Consider the step-index fiber geometry shown in Figure 7. To be coupled into a bound ray, light must enter the fiber and then be totally reflected from the core-cladding interface. From the geometry of Figure 7, it is clear that the angle of the incident ray must be less than ϕ'_c where

$$\sin\theta_c = n_2/n_1 \tag{78}$$

is the sine of the critical angle and the angles ϕ'_c, ϕ_c, and θ_c are related by

$$n_o\sin\phi'_c = n_1\sin\phi_c = n_1(1 - \sin^2\theta_c)^{1/2} \tag{79}$$

The term on the left side of this equation is called the NA of the fiber and can be written

$$NA = (n_1^2 - n_2^2)^{1/2} \tag{80}$$

If $\Delta \ll 1$, the NA for this step-index fiber can be simplified to

$$NA = n_1\sqrt{2\Delta} \tag{81}$$

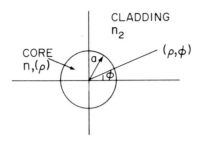

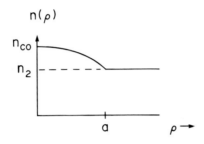

FIGURE 8. (A) Cross section of a graded-index fiber. (B) Variation of the refractive index along the radial direction in a graded-index fiber.

For sources that radiate equally in all directions, the coupling efficiency (i.e., the power coupled into bound rays divided by the power radiated by the source) is proportional to the square of the NA.

It is evident that the NA is not an exact measure of the coupling efficiency. One reason is that both skew rays and leaky rays have been neglected in the calculation of the NA. Both of these will have an influence on the coupling efficiency.[14,15] Another reason is that the radiation pattern of the source is important. Incoherent sources such as LEDs radiate almost uniformly in all forward directions. Thus, for these sources, the coupling efficiency is proportional to the square of the numerical aperture and is usually small. A typical coupling loss in this case is 30 dB. Coherent sources such as solid state lasers, however, have a very narrow radiation pattern centered around the fiber axis. Hence most of the energy is coupled into the bound rays, and the coupling efficiency is relatively good. A typical coupling loss for laser sources is 3 dB. A further reason that NA is not an exact measure of coupling loss is that often the coupling efficiency is increased by using lenses at the end of a fiber. Despite these difficulties, NA is a useful measure of light acceptance and is commonly used.

IX. RAY DISPERSION IN MULTIMODE GRADED-INDEX FIBERS

A remedy for the problem of pulse broadening in step-index fibers can be found by constructing a fiber in which the core refractive index gradually decreases away from the fiber axis. A fiber of this type is called a graded-index fiber and is illustrated in Figure 8.

The reason why this type of fiber exhibits less pulse broadening can be understood in the following way. As in the step-index fiber, the shortest ray path follows the fiber axis, while longer ray paths oscillate around the ray axis as shown in Figure 9. Along most of the oscillating ray path, however, the refractive index is smaller than on the fiber axis. Since the ray velocity is proportional to $1/n$, the light that travels along the oscillating paths travels at a higher speed than the light that follows the axis. This greater speed partially compensates

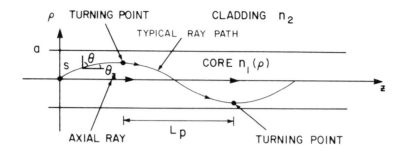

FIGURE 9.   Ray path geometry in a graded-index multimode fiber.

for the longer distance traveled. This result suggests that transit times for the oscillating ray paths are not very different from the transit time for the axial ray path. Since pulse broadening is proportional to the maximum difference in transit times, a smaller pulse broadening can be expected for graded-index fibers. In the following paragraphs, a quantitative measure of pulse broadening for a graded-index fiber will be derived.

In graded-index fibers, ray paths are no longer straight lines as they are for the step-index fiber. The ray trajectories must be determined by solving a pair of differential equations. For a meridional ray

$$\frac{d}{ds}\left[n_1(\rho)\frac{d\rho}{ds}\right] = \frac{dn_1(\rho)}{d\rho} \tag{82}$$

$$\frac{d}{ds}\left[n_1(\rho)\frac{dz}{ds}\right] = 0 \tag{83}$$

where s is the distance along the ray path as shown in Figure 9.[18] The solution to these differential equations is a pair of functions, $\rho(s)$ and $z(s)$, that form a parametric representation for the ray trajectory. Alternatively, the equations can be solved for $z(\rho)$, the explicit trajectory, as will be done below. Equation 83 is integrable and yields

$$n_1(\rho)\cos\theta_z(\rho) = n_1(0)\cos\theta_z(0) \tag{84}$$

where

$$\theta_z = \frac{\pi}{2} - \theta$$

Equation 84 is a generalization of Snell's law of refraction for an inhomogeneous medium. Note that it is written in terms of θ_z, the angle with respect to the fiber axis. Thus, $\cos\theta_z$ appears instead of the more familiar version of Snell's law in which $\sin\theta$ appears.

From Equation 84, a constraint on ray trajectories can be deduced. As the ray moves off the axis, $n_1(\rho)$ decreases. According to Equation 84, $\cos\theta_z$ must increase since the right-hand side is constant. Hence, θ_z decreases and the ray path must flatten out. Since $\cos\theta_z$ can never be larger than 1, the ray can never reach beyond the point for which

$$n_1(\rho_{tp}) = n_1(0)\cos\theta_z(0) \qquad 0 \leq \rho_{tp} \leq a \tag{85}$$

This point ρ_{tp} is called a turning point and is the point at which the ray turns back toward the fiber axis. Rays that have turning points smaller than a are said to be bound rays by analogy to the bound rays of a step-index fiber.

Equation 82 can now be solved for the ray path. The details of the solution are outlined in a homework problem. The result is

$$z(\rho) = n_1(0) \cos \theta_z(0) \int_0^\rho \frac{d\rho}{[n_1^2(\rho) - n_1^2(0) \cos^2 \theta_z(0)]^{1/2}} \tag{86}$$

The length of one half cycle of the ray path will be called L_p. From the symmetry of the ray (as illustrated in Figure 9):

$$L_p = 2z(\rho_{tp}) \tag{87}$$

As mentioned earlier, the optical path length of a ray in a homogeneous medium is the product of the refractive index and the path length. In an inhomogeneous medium, the optical path length can be constructed by summing the product of the refractive index and the path length for each infinitesimal section of path length. The result, for a ray which begins at z = 0 and travels through a length z of fiber is

$$L_o(z) = \int_0^{s(z)} n_1(s)ds \tag{88}$$

Here, again, s is the distance along the ray path measured from z = 0. For a particular refractive index profile $n_1(\rho)$, the optical path length for a half cycle of the ray path is

$$L_o = 2 L_o(z_{tp}) = 2 \int_0^{\rho_{tp}} \frac{n_1^2(\rho) \, dp}{[n_1^2(\rho) - n_1^2(0)\cos^2 \theta_z(0)]^{1/2}} \tag{89}$$

where $z_{tp} = z(\rho_{tp})$ from Equation 86. Note that if $n_1(\rho)$ is a constant, Equation 89 reduces to the optical path length for a homogeneous medium.

In a homogeneous medium, the transit time for a ray traveling between two points is the length of the path divided by the speed at which the ray propagates. The speed, which is constant in a homogeneous medium, is equal to the speed of light divided by the refractive index of the medium. In an inhomogeneous medium, the transit time may be obtained by summing each incremental length divided by the average speed of the ray in that length. In the limit as the incremental length goes to zero

$$T(z) = \frac{1}{c} \int_0^{s(z)} n_1(s)ds \tag{90}$$

Aside from the constant velocity of light, this is identical to the optical path length. Thus the transit time for a ray that travels through a length of L of fiber can be written approximately as

$$T(L) \simeq \frac{L_o L}{c L_p} \tag{91}$$

Strictly speaking, this formula is valid only when the length L is equal to an integer times $L_p/2$. However, if $L \gg L_p$, it is quite accurate even for other distances.

To study the pulse broadening characteristics of graded-index fibers, it is necessary to identify bound ray paths with the smallest and largest transit times. The difference between these two times is the pulse broadening. To be more quantitative about the pulse broadening characteristics of graded-index fibers, the special case of a parabolic refractive index profile will be examined. The refractive index profile is assumed to be

$$n_1(\rho) = n_1(0)\left(1 - 2\Delta\left(\frac{\rho}{a}\right)^2\right)^{1/2} \simeq n_1(0)\left(1 - \Delta\left(\frac{\rho}{a}\right)^2\right)$$

$$\Delta \ll 1 \tag{92}$$

For this profile, the ray paths can be shown to be sinusoidal by integrating Equation 86. The details of this integration will be left to a home-work problem.

The result is

$$\rho(z) = \frac{a(1 - \cos^2 \theta_z(0))^{1/2}}{\sqrt{2\Delta}} \sin\left(\frac{\sqrt{2\Delta}\ z}{a\ \cos\theta_z(0)}\right) \tag{93}$$

The ray half-period can be calculated from Equation 93 to be

$$L_\rho = \frac{\pi a \cos\ \theta_z(0)}{\sqrt{2\Delta}} \tag{94}$$

The optical path length calculated from Equation 89 is

$$L_o = \frac{n_1(0)\ a\pi\ \cos\ \theta_z(0)}{2\sqrt{2\Delta}} \left[\cos\ \theta_z(0) + \frac{1}{\cos\ \theta_z(0)}\right] \tag{95}$$

From Equations (90) and (94), the transit time for a ray can be calculated as

$$T(L) = \frac{n_1(0)L}{2c} \left[\cos\ \theta_z(0) + \frac{1}{\cos\ \theta_z(0)}\right] \tag{96}$$

The ray paths with the smallest and largest transit times are the rays for which $\theta_z(0) = 0$ and $\cos\ \theta_z(0) = (1 - 2\ \Delta)^{1/2}$, respectively. Note that a path for which $\cos\ \theta_z(0) > (1 - 2\Delta)^{1/2}$ will carry the ray beyond $\rho = a$ contrary to the definition of a bound ray. Thus, the pulse broadening is approximately*

$$T_d(L) = \frac{3n_1(0)L\Delta^2}{4c} \tag{97}$$

For this result, it can easily be seen that, for a given value of Δ, the pulse broadening due to ray dispersion is much smaller than that for a step-index fiber. This difference is the reason why essentially all high-speed multimode optical fiber systems use graded-index fibers.

A typical pulse broadening for a graded-index fiber is 3 nsec/km. This is a considerable improvement over the step-index fiber. However, it does not approach the theoretical improvement in pulse broadening predicted by Equation 97. The reason it does not is due to the difficulty in constructing fibers with a well-controlled refractive index profile for a reasonable cost.

It should be noted that the numerical aperture (NA) for a graded-index fiber and a step-index fiber with the same value of Δ is identical.[14] With the reservations noted earlier, NA is a measure of coupling efficiency between a source and a fiber. Thus, for a given coupling efficiency, graded-index fibers have a significantly better pulse distortion performance than step-index fibers.

---

* As in the case for the step-index fiber (Equation 77) this is not RMS pulse broadening. For the parabolic index fiber, the RMS broadening is smaller than that given by (Equation 97) by a factor of 6 $\sqrt{3}$.[12]

It should be mentioned that the parabolic profile is not the only profile that has been considered for graded-index fibers. In fact, it is not the profile that results in the optimum pulse broadening performance; however, it is close to the optimum. A study of alternate fiber profiles can be found in a number of other references.[12,14]

X. RAY COUPLING AND MODAL NOISE EFFECTS

The fact that a ray path is a permissible path for light from the source to the detector does not mean that light will actually follow that path. The amount of light carried along each potential ray path is determined by the particular light source that is used. In ideal fibers, such as the ones that have been discussed here, the rays that are excited by the source continue undisturbed until they reach the detector. This situation does not actually occur for at least two reasons. First, there are random microscopic irregularities that cause scattering of rays away from the direction in which they were originally traveling. Energy scattered into rays that are not bound is lost. This lost energy is the cause of the Rayleigh scattering limit on fiber attenuation that was discussed in Section IV. The scattering also results in a coupling of energy from one bound ray to other bound rays. Second, scattering may be caused by discontinuities in the fiber such as those encountered at splices. This scattering will have effects similar to those described above.

The significance of these factors will now be investigated. First, consider the effect of ray coupling. The result of this is that the original distribution of energy among the rays set up by the source changes as the rays propagate through the fiber. Eventually, the distribution of energy among the rays no longer changes because, for each ray, energy is "scattered in" and "scattered out" at a rate that preserves the relative amplitude of each ray. When this condition exists, it is said that "steady state" has been achieved. The length needed to achieve steady state may be longer than 1 km in high quality fibers.

One consequence of ray coupling is that fiber attenuation and pulse spreading measurements that are made in short fibers are unreliable. This is because the measurement reflects not only the fiber, but also the particular distribution of rays set up by the source. A second consequence of the ray coupling phenomenon is that it causes a decrease in the pulse broadening due to ray dispersion. It was assumed in the derivation of pulse broadening that no energy was transferred from ray to ray. In a real fiber, however, energy is transferred back and forth between rays. Thus the energy alternates back and forth between rays that have short transit times and those that have long transit times. Intuitively, it makes sense that this would cause the differences in transit times to average out and result in a smaller amount of pulse broadening. Marcuse has shown by modal analysis that the pulse broadening is proportional to $\sqrt{L}$ rather than to L as indicated by Equations 77 and 91.[16] Measurements show that the actual pulse broadening falls between these two extremes.[19]

As mentioned above, the distribution of energy among the rays (or modes) depends on the particular source used. Further changes in the ray distribution can be introduced by mechanically stressing the fiber. The reason that this is important is that the reflection and transmission properties of fiber splices depend upon the particular set of rays incident on the splice. This is particularly true if the incident rays have a narrow frequency spectrum because the individual rays can then interfere both constructively and destructively. Thus this phenomenon is usually only observable when single mode lasers are used. The uncertainties in fiber attenuation caused by the source and by mechanical deformation (such as bending) are often grouped into a category called "modal noise".[20]

XI. INFLUENCE OF FIBER PROPERTIES ON SYSTEM PERFORMANCE

In this chapter, the characteristics of optical fibers that influence the propagation of light pulses have been outlined. It is important that the significance of these characteristics relative to system performance be understood. This information will assist the designer in the selection of an optical fiber and associated terminal devices that will satisfy system specifications.

To be consistent with the emphasis of this chapter, only digital communication systems will be considered. The analysis of the performance of these systems will be divided into two parts: attenuation and distortion.*

First, the signal that arrives at the receiver must be of sufficient amplitude to achieve a specified bit error rate. Thus the amount by which the transmitted signal can be attenuated as it travels through the system has an upper limit.

To predict the attenuation, all known sources of attenuation must be identified. These are:

1. Source-fiber coupling loss (dB) — see Section VIII (depends on source type and the fiber numerical aperture)
2. Connector loss (dB)** — typically 1 dB/connection (loss/connection × number of connectors)
3. Splice loss (dB) — typically 0.5 dB/splice (loss/splice × number of splices)
4. Fiber attenuation (dB) — see Section IV (approximately equal to the specified loss/ kilometers × number of kilometers of fiber)
5. Fiber-detector coupling loss (dB) — usually assumed to be zero

The total known system losses can then be added. Given this result, the system must be designed to satisfy the following power balance equation:

[source output power (dBm) − known system losses (specified in 1 to 4 above) − system margin] ≥[minimum required receiver power (for a given bit error rate at given data rate)]

The system margin is necessary due to uncontrollable effects such as temperature, aging, unplanned splices, and modal noise. Typically, the system margin is 6 dB. The power balance equation could be used to solve for the maximum distance between repeaters for a given fiber attenuation or the maximum allowable fiber attenuation for a given system length. It is assumed here that the distortion requirements to be discussed next are satisfied.

The second step in the design process is to insure that fiber dispersion does not cause an unacceptable level of distortion and, hence, intersymbol interference. There are five contributors to the distortion of a pulse as it travels from the input of the optical transmitter to the output of the optical receiver. These are

1. The bandwidth (or rise time) of the optical transmitter
2. Material dispersion of the fiber
3. Ray (or modal) dispersion of the fiber
4. Waveguide dispersion of the fiber
5. The bandwidth (or rise time) of the optical receiver

The bandwidth of the optical transmitter is usually chosen to be much larger than the bit rate of the system. It will be assumed here that this has been done and that the effect of the transmitter on pulse distortion can be ignored. It should be noted that typical solid state lasers have bandwidths of 500 MHz while typical LED bandwidths are a factor of ten smaller. This is one reason that lasers are preferred for high data rate systems.

---

* Actually attenuation and distortion cannot be treated entirely separately. For example, both distortion and attenuation requirements influence the choice of receiver bandwidth. Large bandwidths produce less distortion but increase the signal power required to achieve a given error rate.

** Remateable connections between the fiber and transmitters, receivers and repeaters.

As mentioned earlier, the dispersive characteristics of the fiber will cause distortion of the pulses. This distortion is responsible for intersymbol interference that in turn causes a degradation in the bit error rate of the system. In theory, filtering can be done at the receiver to compensate for this degradation in performance. However, the receiver filters that would be required emphasize high frequencies. If these are used, additional noise is introduced into the receiver that in turn degrades the bit error rate.[14] Due to the uncertainty of significant gains in system performance, compensating filters are not usually used. The more common procedure is to design the system such that the pulse at the receiver input is not significantly distorted. The receiver can then be designed independently of the properties of the particular fiber selected. This is the approach chosen here.

The following paragraphs on pulse broadening have been written specifically for multimode fibers at a wavelength for which waveguide dispersion can be neglected. The ideas, however, can be used in the analysis of monomode fibers with the following changes. First, there is no intermodal pulse broadening. Second, at wavelengths for which material dispersion is small, waveguide dispersion must be considered as in Section V. Finally, the bandwidth of the fiber is related to intramodal dispersion, not intermodal dispersion as assumed here. assumed here.

To quantify the effect of pulse broadening on system performance, the propagation of Gaussian shaped pulses will be considered. The mathematical formula for a Gaussian shaped pulse is

$$h(t) = \frac{1}{\sqrt{2\pi}\,\sigma} \exp\left(-t^2/2\sigma\right) \tag{98}$$

Midwinter has shown that a Gaussian pulse of initial width 0.816 T has a width

$$(\tau_e)^2 \simeq (.816T)^2 + (\phi_m \Delta\lambda L)^2 + (\Psi L^b)^2 \tag{99}$$

after propagating a distance L along an optical fiber.[12] This equation and the variables contained in it will be discussed in the next few paragraphs.

The parameter T is the bit interval [1/(bit rate)] and τ_e is the width of the Gaussian pulse measured between the points for which the pulse has decayed to 1/e of its center value. $\Delta\lambda$ is the RMS spectral width of the source where the RMS width for a function $g(\lambda)$ is defined to be

$$(\Delta\lambda)^2 = \int_{-\infty}^{\infty} \lambda^2 g(\lambda)d\lambda - \left[\int_{-\infty}^{\infty} \lambda g(\lambda)d\lambda\right]^2 \text{ where } \int_{-\infty}^{\infty} g(\lambda)d\lambda = 1 \tag{100}$$

The figure 0.816 is chosen because a Gaussian pulse of this width has the same RMS pulse width as a rectangular pulse of width T. The second and third terms of Equation 99 represent the material and ray contribution to pulse broadening, respectively.

For wavelengths not near the zero material dispersion wavelength, the material dispersion coefficient can be shown to be[14]

$$\phi_m = \left| \frac{2\sqrt{2}}{c} \lambda \frac{d^2n}{d\lambda^2} \right| \tag{101}$$

As above, the pulse width has been measured between the 1/e points of the Gaussian pulse. The value of ϕ_m depends on both the material used for the fiber and the wavelength of the source. For silica glass, the material dispersion coefficient is 240 psec/km-nm at a wavelength of 850 nm and is approximately zero at 1250 nm (the zero material dispersion wavelength).

The last term is related to the optical fiber bandwidth as specified by the fiber manufacturer.* According to Midwinter,[12] τ_e is related to the 3-dB optical bandwidth (BW) of a fiber by

$$\Psi = .53/BW \tag{102}$$

As mentioned earlier, the value of b is somewhere between 0.5 and 1 due to ray coupling. It has been found that for the design of high data rate telecommunications systems, a figure of b = 0.8 is reasonable.

Midwinter has shown that if $\tau_e \leq 1.6T$, then there is less than 1 dB penalty in receiver sensitivity for a given bit error rate. This criteria can be used to insure that dispersion will not be a limiting factor in system performance.

PROBLEMS

1. Derive Equation 34 using Equation 33 and the definitions of ω, β, λ and v_p.
2. Calculate the cutoff frequency for the TE_{01} mode for the fiber of Figure 3 if a = 5 μm.
3. Referring to Figure 3, list the modes which will propagate in the fiber if a = 1.0 μm and λ = 1.4 μm.
4. Derive Equation 5 by using Equation 3 and 4 and the Laplacian in cylindrical coordinates.
5. Starting with Equation 38, derive Equation 39.
6. Derive Equation 70 using Equation 51.
7. Show that Equation 53 reduces to Equation 41 if $n_1 = n_2$.
8. Referring to Reference 9 to obtain needed information about "Sample 2" material:

    (1) List the constants $(A_i, 1_i)$ for i = 1 to 3 from Table 1.[9]
    (2) Calculate the group index N for Sample 2 material, at λ = 1 μm.
    (3) Find the percentage difference between n and N for Sample 2 material at λ = 1 μm.

9. Derive Equation 86 from Equation 82 and Equation 84. Hint: first rewrite the derivatives with respect to s as derivatives with respect to z. Then use the identity F dF/dy = [d(F^2)/dy]/2. Next, use the change of variable y' = dy/dz such that y'(dy'/dy) = d^2y/dz^2. Finally, integrate both sides from y to y_{tp} and then from 0 to y.
10. Derive Equation 93 from Equation 86 and the first form of Equation 92. Use the result

$$\int \frac{dx}{\sqrt{a^2 - x^2}} = \arcsin\left(\frac{x}{|a|}\right)$$

---

*   It is assumed here that the measurement of bandwidth in multimode fibers is done with a source that has a narrow enough linewidth that material dispersion contributions to the bandwidth measurement can be neglected.

REFERENCES

1. **Marcuse, D.,** *Light Transmission Optics,* Van Nostrand Reinhold, New York, 1972.
2. **Keck, D. B.,** Optical fiber waveguides, in *Fundamentals of Optical Fiber Communications,* 2nd ed., Barnoski, M. K., Ed., Academic Press, New York, 1981.
3. **Bell, W. W.,** *Special Functions for Scientists and Engineers,* Van Nostrand, London 1968.
4. **Jenkins, R. A. and White, H. E.,** *Fundamentals of Optics,* McGraw-Hill, New York, 1976.
5. **Gloge, D.,** Dispersion in weakly guiding fibers, *Appl. Opt.,* 10, 2442, 1971.
6. **Jurgensen, K.,** Dispersion minimum of monomode fibers, *Appl. Opt,* 18, 1259, 1979.
7. **Kapron, F. P.,** Maximum information capacity of fiber-optic waveguides, *Electron. Lett.,* 13, 96, 1977.
8. **Chang, C. T.,** Minimum dispersion at 1.55 μm for single-mode step-index fibers, *Electron. Lett.,* 15, 765, 1979.
9. **Pires, P. S. M., Rogers, D. A., Bochove, E., and Souza, R. F.,** Prediction of laser wavelength for minimum total dispersion in single-mode step-index fibers, *IEEE Trans. Microwave Theor. Tech.,* 30, 131, 1982.
10. **Chang, C. T.,** Minimum dispersion in a single-mode step-index optical fiber, *Appl. Opt.,* 18, 2516, 1979.
11. **Miller, S. E., Marcatili, E. A. J., and Li., T.,** Research toward optical-fiber transmission systems, *Proc. IEEE,* 61, 1703, 1973.
12. **Midwinter, J. E.,** *Optical Fibers for Transmission,* John Wiley & Sons, New York, 1979.
13. **Felsen, L. B. and Shin, S. Y.,** Rays, beams and modes pertaining to the excitation of dielectric waveguides, *IEEE Trans., Microwave Theor. Tech.,* 23, 150, 1975.
14. **Miller, S. E. and Chynoweth, A. G.,** *Optical Fiber Telecommunications,* Academic Press, New York, 1979.
15. **Gallawa, R. L.,** A Users Manual for Optical Waveguide Communications, Rep. 0T 76-83, U.S. Department of Commerce, Washington, D.C., March 1976.
16. **Marcuse, D.,** *Theory of Dielectric Optical Waveguides,* Academic Press, New York, 1974.
17. **Snyder, A. W.,** Leaky-ray theory of optical waveguides of circular cross section, *Appl. Phys.,* 4, 273, 1978.
18. **Born, M. and Wolf, E.,** *Principles of Optics,* Pergamon Press, New York, 1975.
19. **Bunker, N. S., Cheung, H. W., and Ester, G. W.,** Design considerations for a 44.7 Mb/sec fiber optic transmission system, Paper No. 24/3, 1980 IEEE ELECTRO Conf., Boston, 1980.
20. **Epworth, R. E.,** The phenomenon of modal noise in analogue and digital fibre systems, Proc. 4th Eur. Conf. Optical Communications, Genoa, 1979, 492.

Chapter 4

OPTICAL SOURCES

Gabriel Lengyel

TABLE OF CONTENTS

I. INTRODUCTION

The rapid expansion of the field of optical communications in the last 20 years was made possible by the development of low-loss optical fibers and of light sources that are compatible with the spectral transmission characteristics and physical dimensions of these fibers. The extremely small diameter of the fibers (a few microns for monomode and 50 to 200 μm for multimode fibers) required light sources of comparable dimensions emitting in the near infrared region (around 1 μm wavelength) where these fibers display a relatively high optical transmittance. Such sources were provided by injection luminescent diodes and somewhat later, by laser diodes that were fabricated from single crystals of III-V compound semiconductors or their ternary or quaternary mixtures. These devices not only provide a good match in physical size and emission wavelength to the fibers, but allow the simple method of modulating the output light through variation of the pumping current. Sources used today in practical optical communication systems belong to this group of III-V semiconductors: GaAs and GaAlAs for the 0.7 to 0.9 μm spectral region and InP with InGaAsP for the 1.3 to 1.6 μm region.

Electroluminescent diodes and lasers can also be fabricated from other semiconductors. II-VI compounds and their mixtures have actually been used for various special applications, but the optical communications field has restricted itself until now almost exclusively to the above mentioned III-V compound structures. Therefore, the discussion of this chapter will refer only to these. Optical communications in the 0.7 to 0.9 μm region can already be regarded as a practical reality while the 1.3 to 1.6 μm range is still in the experimental stage. Most of the discussion in this chapter will refer to the 0.7 to 0.9 μm region. There is no doubt, however, that the longer wavelength region will also become practical in the next few years, particularly for more demanding applications, because optical absorption losses in the fibers are even lower in that spectral region.

It is not the objective of the author to give a complete description of infrared emitting diodes (IREDs) and laser diodes. A full discussion of their underlying theory, design, and manufacturing technology exceeds the scope of this book. A broad treatment of this rather complicated subject is probably not even necessary when one tries to address that larger group of communications engineers and technicians who are more interested in the application of these devices than in their design or development. The more interested reader is referred to the excellent literature already available on the subject.[1-3] Rather, this chapter will focus on those characteristics of these light sources that are important from the applications point of view. Some theoretical background in a concise form will be given where it is felt to be of use for clearer understanding.

The physical characteristics of a light source that are important in optical communication systems are wavelength of maximum emission, spectral bandwidth, light-current characteristics, far-field pattern, dynamic response, polarization and coherence properties, and noise. To these strictly physical parameters one has to add some practical requirements, such as stability against variations in temperature or other ambient parameters, aging characteristics, and economic considerations. The significance of these from the communications point of view will be discussed.

Although the basic light generation mechanism, the radiative recombination of electrons with holes in a forward biased p-n junction is essentially the same in IREDs and lasers, there are enough structural and other differences between these two to warrant a separate treatment in this chapter. IREDs will be discussed first because of their greater simplicity, when compared with lasers. In each case, a brief theoretical background will be given to facilitate the understanding of the operation of the device; then their structure will be discussed.

Laser diodes that can operate in the CW mode at room temperature without an external cooling device are always of the double-heterostructure (DH) type. This structure provides confinement of the injected carriers through a heterojunction and at the same time assures optical guidance of the electromagnetic (EM) wave perpendicularly to the p-n junction (transverse direction). Many different structures have been developed in recent years to provide electrical and optical confinement in the lateral direction, parallel to the p-n junction plane.*

DH lasers can be divided into two large classes according to the method chosen for the lateral guidance of the EM wave. In the so-called gain-guided structures, optical guidance is provided only by the optical gain in heavily pumped regions of the laser, and, therefore, the guidance is current dependent. In the so-called index-guided laser, a discontinuous change in the refractive index is built into the structure in the lateral direction, similar to the refractive index steps in the transverse direction that is characteristic of all DH lasers. Some operating characteristics of gain- and index-guided lasers are quite different; therefore they will be discussed separately in this chapter.

II. THEORETICAL BACKGROUND

The physical processes governing the operation of light or infrared emitting diodes are both optical and electronic in nature. A brief review of the physical background of these processes will be given in order to facilitate the detailed discussion of the devices themselves.

A. Optical Processes

The interaction of electromagnetic radiation with a solid can be characterized on the macroscopic scale by the complex refractive index[4]

$$\bar{n}^* = \bar{n} - i\bar{k} \tag{1}**$$

The real part $\bar{n}$ represents polarization processes in which the average exchange of energy between the EM field and the solid is zero. The imaginary part $\bar{k}$, the extinction coefficient, characterizes interactions in which an energy exchange takes place between the material and the radiation field. This exchange can either be a loss when energy from the field is absorbed by the material ($\bar{k} > 0$) or a gain ($\bar{k} < 0$) when the material contributes photons to the field. This latter process is of great importance for radiation sources, particularly lasers.

The imaginary part of the refractive index describes how an EM wave is attenuated or amplified on passing through the material. In optics, the most accessible quantity to measurements is the intensity of the wave and not its amplitude; therefore, the attenuation coefficient refers usually to the attenuation of intensity. It is related to $\bar{k}$ as

$$\alpha = 4\pi\bar{k}/\lambda_o \tag{2}$$

where λ_o is the wavelength in a vacuum. In an optically active medium in which the wave is amplified, $\bar{k}$ becomes negative and the symbol g is often used instead of $\bar{k}$.

The relationship between the complex relative dielectric constant and the complex refractive index is given by

---

* The nomenclature follows the usage of the literature: the direction of wave propagation (optical axis) is called longitudinal, the direction perpendicular to the p-n junction and the optical axis is referred to as transverse, and the direction parallel to the p-n junction and perpendicular to the optical axis, lateral.

** The bars over the real and imaginary parts are often used in the literature to avoid confusion with the electron density n and the wave number k.

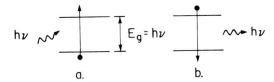

FIGURE 1.   Absorption and emission process: (a) absorption; (b) emission.

$$\epsilon_{rel}^* = \epsilon' - i\epsilon'' = (\bar{n}^*)^2$$

$$\epsilon' = \bar{n}^2 - \bar{k}^2$$

$$\epsilon'' = \pm\, 2\overline{nk} \tag{3}$$

Finally, one has to add that all optical constants are wavelength dependent, and strong variations occur particularly in that region of the spectrum in which the device emits. The real and imaginary parts of the complex optical constants are not independent from each other; they are linked by the Kramers-Kronig relations.[4]

Considering the energy exchange processes between a solid and optical radiation, one has to return to Einstein's theory of the photoelectric effect (Figure 1). In the absorption process, a photon raises the energy of an electron by the amount of its own enegy; in the emission process, the opposite takes place. In absorption, the photon is annihilated; in emission, a photon is created. This latter process forms the basis of all light-emitting devices.

In a semiconductor light-emitting device, the top of the valence band represents the lower energy level where most of the electrons are under the conditions of thermal equilibrium. The bottom of the conduction band becomes the upper energy level. Thus the light emitted corresponds more or less to the energy gap of the semiconductor

$$E_g = h\nu = hc/\lambda_o \tag{4}$$

Practical light-emitting devices are made of heavily doped semiconductors. In these materials, several different kinds of recombination mechanisms are possible[5] (Figure 2). The illustrated processes are not equally probable; furthermore, not all recombinations produce an emittable photon. Auger processes (Figure 2d), for example, are always nonradiative. In that, and other nonradiative processes, the energy of recombination is transmitted to the lattice and is lost for light generation. An important example for nonradiative recombination takes place on crystal surfaces and interfaces. Such recombinations reduce the efficiency of the device and contribute to undesired heat losses. For light generation, the recombinations under a and b in Figure 2 are of the greatest importance. Which transitions will dominate the recombination process depends on the band structure of the material.

Figure 3 shows the band structure of GaAs and GaP in momentum ($\vec{k}$) space. In a direct bandgap material such as GaAs, the minimum bandgap occurs at $\vec{k} = 0$ where radiative transitions are most likely. In indirect bandgap materials (e.g., GaP), transitions between the conduction band minimum and valence band maximum require the assistance of a photon ($\hbar\omega$) because the momentum of the system must be conserved ($\vec{k}$ selection rule). Such three-particle processes are less likely. Indirect recombinations can also be assisted by properly chosen impurity atoms as it is frequently done in GaP LEDs.

In a semiconductor crystal, the rate of generation of hole-electron pairs by absorption of photons from the surrounding radiation field is equal to the rate of recombination when the crystal is in thermal equilibrium with its surroundings. If an increased rate of recombination

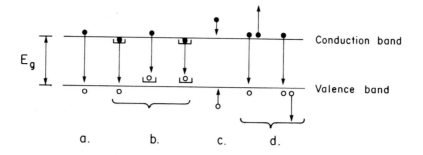

FIGURE 2.   Recombination processes: (a) band-to-band; (b) recombinations involving impurity sites; (c) intraband recombinations; (d) Auger process.

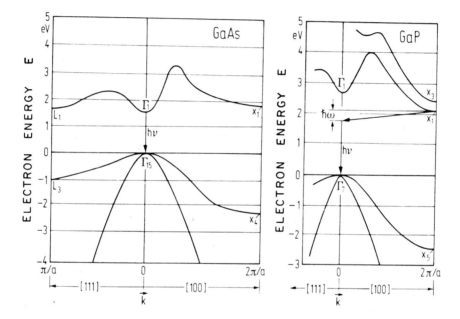

FIGURE 3.   Band structure of GaAs and GaP at 0°K. Arrows show band-to-band recombinations in a direct bandgap (GaAs) and an indirect bandgap (GaP) material; $\vec{k}$: electron momentum; a: lattice constant. (After Winstel, G. and Weyrich, C., *Optoelektronik I*, Springer-Verlag, Basel, 1980. With permission.)

is required as in light emitting devices, additional carriers have to be brought to the conduction band by some external excitation. This excitation or pumping is provided in IREDs and laser diodes by the injection of carriers through a forward biased p-n junction. If the pumping is stopped, these excess carriers recombine spontaneously and disappear exponentially with time [$\exp(-t/\tau)$] where τ is the spontaneous recombination lifetime. τ is of the order of a a few nanoseconds in GaAs and GaAlAs and is an important parameter determining the dynamic behavior of the device. To distinguish between radiative and nonradiative processes, one has to attach different lifetimes to each of these:

$$\tau^{-1} = \tau_r^{-1} + \tau_{nr}^{-1} \tag{5}$$

The internal quantum efficiency of the material is defined in terms of the recombination rates as

$$\eta_i = R_r/(R_r + R_{nr}) \tag{6}$$

where R_r is the radiative recombination rate, R_{nr} is the nonradiative recombination rate (in $cm^{-3}sec^{-1}$). Here, the recombination rates are proportional to the excess carrier density and inversely proportional to the respective lifetimes. The number of photons created per unit volume and time is equal to R_r. The internal quantum efficiency increases with pumping and approaches unity in GaAs/GaAlAs laser diodes.

In the general case, the total spontaneous recombination rate $R = R_r + R_{nr}$ is proportional to the product of electrons and holes present

$$R = B(p_o + \Delta p)(n_o + \Delta n) \tag{7}$$

where p_o, n_o are the equilibrium carrier densities and Δp, Δn are the excess carrier densities. The recombination coefficient B has an approximate value of 10^{-10} cm^3/sec in GaAs. At the injection levels encountered in practical devices, particularly lasers, the term $Bp_o n_o$ is small compared to the others. Thus, if the semiconductor is heavily doped, $p_o >> n_o$ or $n_o >> p_o$ Equation 7 becomes

$$R \approx B(p_o + \Delta p)\Delta n$$

or

$$R \approx B(n_o + \Delta n)\Delta p \tag{8}$$

The recombination lifetime is

$$\tau = \frac{\Delta n}{R} = [B(p_o + \Delta p)]^{-1} \approx (p_o B)^{-1}$$

or

$$\tau = \frac{\Delta p}{R} = [B(n_o + \Delta n)]^{-1} \approx (n_o B)^{-1} \tag{9}$$

If the injection level is low compared to the background doping, the recombination lifetime is independent of the injection, and the recombination rate becomes a linear function of the excess carrier density. $R \cong \Delta n/\tau$ or $R \cong \Delta p/\tau$.

If the semiconductor is lightly doped or the injection level greatly exceeds the background concentration, one has to deal with bimolecular recombination where the recombination rate is not a linear function of the excess carrier density. In the limit $\Delta n \approx \Delta p$ and

$$R = B\Delta n^2 \tag{10}$$

A lifetime, strictly speaking, cannot be defined in this case because of the nonlinearity of the bimolecular recombination law. For small changes however, the law can be linearized, and the effective lifetime becomes a function of the injection level.

In lasers one has to deal also with an additional recombination mechanism, namely the stimulated recombination. According to Einstein's theory of radiation, the probability of a photon being emitted in unit time is not a constant corresponding to the spontaneous process, but increases linearly with the number of photons present whose energy is equal to the energy of the radiative transition. This positive feedback mechanism in light generation forms the basis of all laser oscillators. The stimulated recombination rate becomes significant only at very high injection levels ($\Delta n \approx 2$ to 4×10^{18} cm^{-3} in GaAs). The stimulated recombination rate is then proportional to the flux $S(cm^{-2}sec^{-1})$ of those photons which have the required energy.

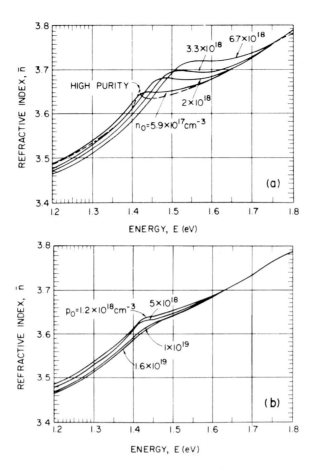

FIGURE 4. Refractive index of GaAs at 297K; (a) n-type; (b) p-type.[6] (After Casey, H. C. and Panish, M. B., *Heterostructure Lasers,* (Parts A and B), Academic Press, New York, 1978. With permission.)

Finally, some numerical values of the most important optical constants of GaAs are of interest. Figure 4 shows the real part of the refractive index ñ for n-1 and p-doped GaAs in the region of photon energies pertinent to device operation (E_g = 1.53 eV, $\lambda_o \cong 0.86$ μm). The refractive index of $Ga_{1-x} Al_xAs$ as a function of the mole concentration x is shown in Figure 5. Note the decrease of ñ with increasing Al concentration. This effect is used in lasers to produce waveguide structures. Replacing part of the gallium by aluminum also increases the bandgap and provides a means, therefore, to reduce the emission wavelength. The empirical relationship between the bandgap energy and the mole concentration of Al in GaAlAs can be given as:[8]

$$E_g = 1.424 + 1.266x + 0.266x^2 \text{ (eV)} \tag{11}$$

This equation is valid between $0 < x < 0.37$. Beyond this range GaAlAs becomes an indirect bandgap material. The absorption coefficient of GaAs is shown in Figure 6. Below the bandgap energy (1.43 eV), the absorption decreases nearly exponentially for the heavily doped samples shown in the figure. If the material is pumped by the injection of carriers, the absorption coefficient changes sign in the vicinity of the bandgap energy. Calculated gain curves for GaAs are shown in Figure 7. The spectrum of the luminescent emission at

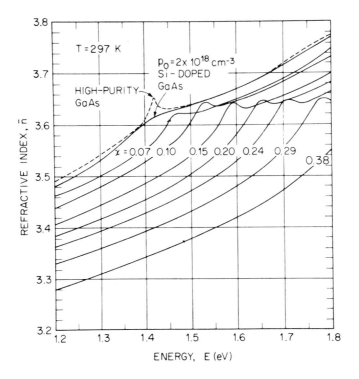

FIGURE 5.   Refractive index of $Ga_{1-x} Al_x As$ at 297K.[7] (After Casey, H. C. and Panish, M. B., *Heterostructure Lasers,* (Parts A and B), Academic Press, New York, 1978. With permission.)

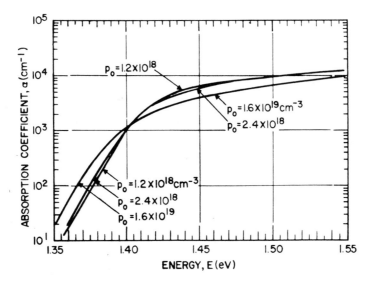

FIGURE 6.   Absorption coefficient of GaAs.[9] (After Casey, H. C. and Panish, M. B., *Heterostructure Lasers* (Parts A and B), Academic Press, New York, 1978. With permission.)

room temperature is given in Figure 8. This spectrum corresponds closely to that of an electroluminescent diode.

B. Electronic Processes

In electroluminescent diodes and diode lasers, the needed excess electron density in the

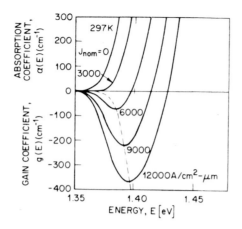

FIGURE 7. Calculated gain curves of GaAs as a function of the photon energy and excitation.[10] J: the pumping rate. (After Casey, H. C. and Panish, M. B., *Heterostructure Lasers,* (Parts A and B), Academic Press, New York, 1978. With permission.)

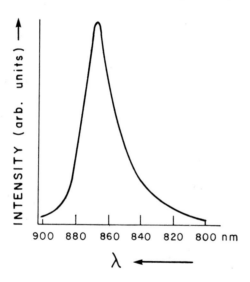

FIGURE 8. Luminescent spectrum of GaAs at 297K.[11]

conduction band is provided by the injection of carriers into the recombination (active) region of the diode through a forward biased p-n junction. Thus the electronic transport properties of the materials used in these devices are just as important as the optical characteristics. It is assumed that the reader is familiar with the fundamentals of p-n junction theory.[12] Current through a forward biased p-n junction follows an exponential dependence on voltage:

$$I = I_o[\exp(eV/kT) - 1] \approx I_o \exp(eV/kT) \tag{12}$$

with V the applied forward bias and k Boltzmann's constant. In practice, the e/kT factor in the exponential must be multiplied by an empirical constant of approximately 0.5 to 1 in order to take into account certain deviations from the idealized theoretical case. At the high injection levels encountered in laser diodes, the Maxwell-Boltzmann statistics used in Equation 12 have to be modified by the more accurate Fermi-Dirac statistics. The exponential relationship remains a good approximation, nevertheless, for a wide range of operating currents. Another important quantity is the injected carrier density that also bears an exponential relationship to the effective barrier height of the p-n junction. Thus the electron density injected into a p-material, using Maxwell-Boltzmann statistics, is

$$n_p = n_n \exp[-e(V_{bi} - V)/kT] \tag{13}$$

where n_n is the electron density on the n-side of the junction, V_{bi} the difference in Fermi levels in the n- and p-materials and V the applied forward bias. A similar relationship holds for holes. From these relations, it is evident that high injection levels require heavy doping (large n_n) and forward bias voltages which can substantially reduce V_{bi}. ($V_{bi} \approx 1.4$ V for GaAs).

Carriers injected across a junction tend to diffuse away from it while they recombine with carriers of the opposite polarity. This diffusion process is best characterized by the diffusion length

$$L = (\tau D)^{1/2} \tag{14}$$

with D the diffusion constant and τ the recombination lifetime. The diffusion length of electrons in GaAs is of the order of a few μm. This means that the number of injected carriers diminishes greatly a few μm away from the junction, and most of the light generation must necessarily take place in the immediate vicinity of the junction. In order to increase the efficiency of the device and reduce pumping current requirements, one has to obtain high carrier densities concentrated into a small volume. From this point of view, diffusion that tends to smear out carrier distributions is an undesirable process with which one, nevertheless, has to live.

One way of preventing the diffusion of carriers is the use of heterojunctions. This not only provides an effective barrier against diffusion, it also improves injection efficiency and, in the case of lasers, serves as an effective guide for the EM wave. An indispensable condition for the application of heterojunctions is a very good lattice match between the two crystals on both sides of the barrier. Lattice mismatch gives rise to defects near the interface that reduce the radiative quantum efficiency and also acts as traps for carriers. The required good match is ensured between GaAs and $Ga_{1-x}Al_xAs$ when x is not too large and between InP and InGaAsP within a limited composition range.

The theory of such ideal, abrupt heterojunctions was worked out by Anderson.[13] The difference in energy gaps between the two terminals produces discontinuous jumps (ΔE_c, ΔE_v) in the energy level diagrams of the junctions (Figure 9). With forward bias, holes can easily be injected from the p-GaAlAs into the n-GaAs while the barrier ΔE_c prevents the escape of electrons from the GaAs. Thus injection efficiency improves. Placing another n-n heterobarrier on the other side of the GaAs (Figure 10), prevents the injected holes from diffusing too far away from the junction as ΔE_v which is of the order of a few tenths of an eV, blocks their way. Thus the recombination region is confined between the two heterojunctions. This principle is used in both lasers and in heterojunction IREDs.

III. PHYSICAL CHARACTERISTICS OF LIGHT SOURCES PERTINENT TO OPTICAL COMMUNICATIONS SYSTEMS

The characteristics considered here refer mainly to external characteristics. Such a characterization is not such a simple task as may appear at first glance, because it involves many parameters that have some bearing on the performance of an optical communications system.

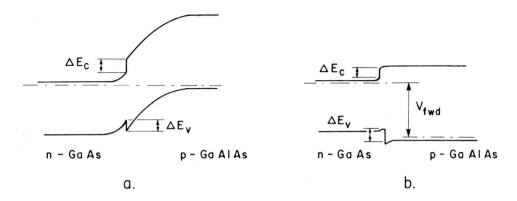

FIGURE 9. Heterojunction between GaAs-n and GaAlAs-p. (a) zero bias; (b) forward bias.

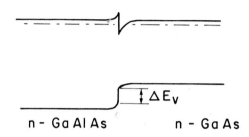

FIGURE 10. Heterojunction between GaAs-n and GaAlAs-n.

A. Spectral Characteristics

Both IREDs and lasers operate in a relatively narrow spectrum. The wavelength of maximum intensity is determined primarily by the bandgap of the material in which the injected carriers recombine. There are other factors involved as well, such as optical absorption that tends to push the emission peak to somewhat longer wavelengths. Optical absorption at the bandgap radiation is high; thus photons of somewhat lower energy become less absorbed on their way out from the device. The doping of the materials has also some influence on the spectrum through the position of the Fermi level. GaAs and GaAlAs devices operate in the 0.7 to 0.9 µm region; InGaAsP in the 1.3 to 1.6 µm region. As the transmission characteristics of fibers (absorption, dispersion, scattering) and the sensitivity of detectors are all wavelength dependent, the importance of the emission wavelength becomes immediately evident.

There is a significant difference in spectral bandwidth between IREDs and lasers. Both are relatively narrow compared to conventional light sources, but laser spectra are much narrower than those of electroluminescent diodes. The spectral bandwidth of a 3000 K black body is approximately 1.5 µm (15,000 Å) with a peak at about 1 µm; for a GaAs IRED this width reduces to about 350 to 400 Å with a peak near 0.9 µm (Figure 11A). In the case of lasers, one has to distinguish between the so-called gain-guided and index-guided structures. Gain-guided lasers oscillate in several longitudinal modes simultaneously, which yield a half-power bandwidth of about 30 to 40 Å. (Figure 11B). Index-guided lasers have an almost monomode spectrum the width of which is less than 0.2 Å at room temperature (Figure 11C). Values of the order of 10^{-4} Å can be reached at low temperatures.[14]

The spectral bandwidth of the source affects the communication system through dispersion in the fibers. Dispersion reduces the distance range of pulse code modulation (PCM) systems or other modulation schemes in which the phase of the wave at the receiver carries information. Photons of different wavelength propagate at different velocities in the fiber. This

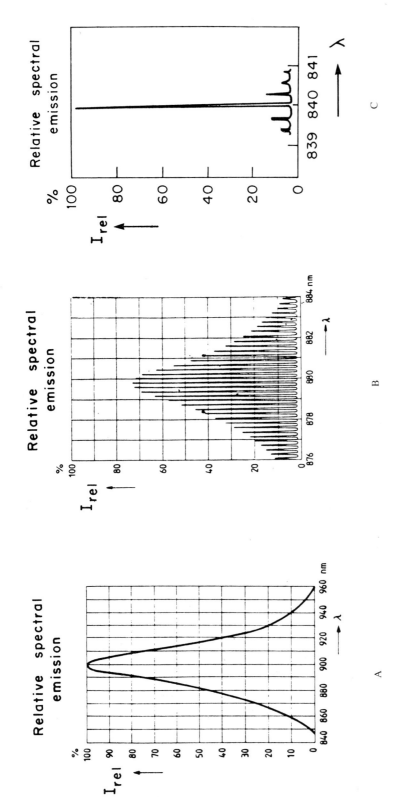

FIGURE 11.   Spectra of typical sources. (A) IRED; (B) multimode laser; (C) monomode laser.[16,17]

result in pulse broadening or varying phase shifts. This effect is proportional to the spectral bandwidth; therefore, a narrow spectral bandwidth is a desirable feature of the source. Extremely narrow bandwidths have disadvantages as well. Modal noise that is a result of speckle patterns is reduced when broad-band, that is, less coherent sources are used. Sources which have very narrow (≈ 0.2 Å) spectral bandwidth are also very sensitive to optical feedback.

B. Light-Current Characteristics

Electroluminescent and laser diodes are current driven devices having nearly linear light output-current characteristics. In the case of laser diodes, one has to distinguish between two fundamentally different regions (Figures 12B and C), one below the lasing threshold and one above it. In both regions, the characteristics are nearly linear connected by a sharply changing curved region near the lasing threshold. The slope of the characteristic is small below threshold and increases dramatically once lasing is obtained.

A key number characterizing the device is the total output power in CW operation that varies from a few 100 μw to a few milliwatts for luminescent diodes and from a few milliwatts to a few 10 mW for communications type laser diodes. Outputs in the 100 mW range may become possible in the not too distant future. Before one attaches too much significance to these numbers, however, one has to consider the fact that in communications applications only that part of the output power is meaningful which can be usefully coupled into the optical fiber. This also depends on the far-field emission pattern of the emitter, and different types of diodes greatly differ from each other on this point. Far-field characteristics will be discussed in more detail in Section C below. The power coupled into the fiber determines the signal to noise figure of the system.

The slope of the light output-current characteristic is a measure of the overall quantum efficiency of the device. In addition to the inherent quantum efficiency of the materials, it includes geometrical and other design parameters. Values range from 0.01 to 0.02 W/A for 0.8 to 0.9 μm. IREDs and 0.2 to 0.4 W/A for laser diodes once they operate in the lasing region. This figure, also called the differential quantum efficiency is lower (0.15 to 0.2 W/A) for gain-guided structures, but can be as high as 0.4 W/A for some index-guided structures. In comparing these figures with the theoretically obtainable value of 1.4 W/A for GaAs, one has to keep in mind that the quoted values refer to output power only on one side of the laser, the customary form of presenting laser data; furthermore, other design considerations do not allow the emission of the entire power present in the laser cavity.

The slopes of the light-current characteristics have a practical significance in the design of drive circuits. They determine the current requirements for biasing and modulation. Laser diodes are biased to or just above threshold in PCM systems and somewhere near the middle of the lasing characteristic in analog modulation schemes. The voltage across the terminals of the device has a logarithmic relationship to the current as discussed in Section IIA. A slight deviation from this theoretical dependence is due to series resistance effects: contact resistance, resistance between contacts, and the junction region of the diode. The saturation value of the voltage lies a few tenths of a volt above the value corresponding to the bandgap for laser diodes and a few tenths below for IREDs. With these figures the required bias and modulation power can be estimated. As lasers operate in the saturated region of the V-I characteristic, the change in voltage during modulation is relatively small.

The linearity of the characteristics is of great importance in analog modulation schemes. Deviations from linearity cause nonlinear distortions of the modulating waveform. In multichannel analog schemes, modulation cross-products caused by nonlinearities result in cross talk between channels. IREDs have generally linear characteristics except some high-power, super-radiant types whose characteristics display a marked super-linear tendency. Characteristics of lasers are highly linear provided one stays entirely in the lasing region. Still some

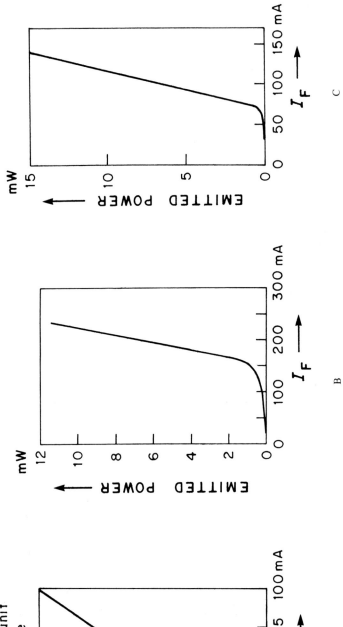

FIGURE 12.   Typical light-current characteristics in cw at room temperature; (A) IRED; (B) gain-guided laser; (C) index-guided laser.[16,17]

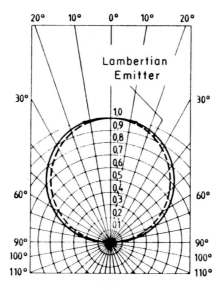

FIGURE 13. Polar radiation pattern of IRED. The dotted circle corresponds to the ideal Lambertian emitter. (Courtesy of Siemens Corporation, Semiconductor Components for Optical Communications. Data book of Optoelectronic Semiconductors, Munich, 1980/81. With permission.)

nonlinear effects can be measured:[18] second harmonic distortions lie about 40 to 60 dB and third harmonic distortions 60 to 80 dB below the fundamental. The lower distortion figures apply to index-guided, the higher ones to gain-guided structures.

C. Far-Field Emission Patterns

Electroluminescent diodes that do not have integral focusing elements and operate as plain surface emitters have a polar radiation characteristic that follows a simple cosine law (Figure 13). Thus they resemble an ideal Lambertian radiator. This means that the power which can be coupled into an optical fiber is limited to a few percent unless one uses some focusing optics between diode and fiber. Without such optics the coupling efficiency is approximately

$$\eta_c \approx \sin^2 \theta_c \approx (NA)^2 \tag{15}$$

with θ_c the critical acceptance angle and NA the numerical aperture of the fiber. Edge emitting IREDs have asymmetrical emission patterns resembling those of laser diodes.

Far-field patterns of laser diodes are much more complicated. One is accustomed from gas lasers to measure beam divergence in milliradians; in the case of laser diodes these numbers increase to a few tenths of a radian. This large divergence is due to the high refractive index of the laser material (3.5 to 3.6) and the extremely small size of the laser cavity. In a coarse approximation, one can regard this phenomenon as the diffraction of light through a small opening. The size of the opening corresponds to the cross section of the laser cavity that measures about 0.1 to 0.2 μm perpendicular to the p-n junction and about 3 to 10 μm parallel to it. In addition to the geometry of the waveguide, other factors heavily involved in determining beam divergence are the optical constants of the materials forming the waveguide. The large difference in the two dimensions of the pinhole is responsible for the widely differing beam divergence measured parallel and perpendicularly to the junction plane. The divergence is much larger perpendicular to the junction. The lack of cylindrical symmetry around the optical axis (astigmatism) is particularly unpleasant in fiber optic communications because the fiber with its cylindrical symmetry calls for a cylindrical radiation pattern. One is often forced, therefore, to use cylindrical lenses in the focusing optics to reduce the asymmetry in the focused spot. Furthermore, lasers often produce complicated phase fronts that are neither planar nor spherical.

There are significant differences between gain- and index-guided lasers in their far-field radiation patterns (Figure 14). These figures refer to lasers that oscillate in the fundamental transverse and lateral modes. The two side lobes parallel to the p-n junction in the gain-guided laser are not due to higher order lateral modes, but to the complicated lateral distribution of the complex refractive index inside the laser that is a characteristic feature of this laser type. Index-guided structures have the advantage of possessing a lower degree of astigmatism.

It is impossible to give simple formulas for the coupling coefficient between a laser and a fiber. This depends on the far-field pattern of the laser, the characteristics of the fiber (diameter, acceptance angle, mono- or multimode), and the focusing optics. The coupling loss for a well designed system amounts to about -2 to -5 dB.

D. Dynamic Characteristics

An attractive feature of luminescent and laser diodes is the ease with which their output can be modulated. The variation of the driving current gives nearly proportional changes in light output. The speed with which the diode responds to a change in drive depends on the electrical characteristics of the driving circuit and the particular carrier recombination mechanism involved in the generation of light. The first will not be discussed here in detail. These devices, being extremely small, have small capacitances (in the order of a few pF), and short leads can reduce series inductance effects. Mounting the device on an integrated drive-circuit-chip reduces stray inductance and capacitance even further. The intrinsic limitations in speed of response are given by the recombination lifetimes. In electroluminescent diodes this refers to the spontaneous and in laser diodes to stimulated recombination lifetimes.

The frequency-response of luminescent diodes follows the features of a low pass filter (Figure 15).[19] The recombination rate, not being completely proportional to excess carrier density, can be increased and the time constant decreased with increased pumping current. Thus, modulation rates in the 100 MHz range can be attained.

The recombination time constant for stimulated processes in laser diodes is shorter than the spontaneous recombination time constant by about an order of magnitude; thus modulation frequencies in the low GHz range can be reached with them. To keep the diode operating in the stimulated (lasing) region, the bias current has to be chosen appropriately, corresponding to the modulation scheme (PCM or analog).

The dynamic response of a laser diode can not be described as simply as that of a luminescent diode. There are several nonlinear processes involved, e.g., the dependence of the recombination rate, and with this that of the optical gain, on carrier density; furthermore, the conservation equations written for carriers and photons lead to coupled nonlinear differential equations that can be solved only numerically (Section V). Figure 16 shows the response of the carrier density (n) and light output (S) to a step-like current drive for lasers having resonators with different optical losses. Lower losses lead to transient oscillations that lengthen the time for the laser to reach its steady state output. Gain-guided lasers have optical resonators with high losses (larger β values); they resemble case (c) in Figure 16. Frequency response curves of typical laser diodes are given in Figure 17.

Finally, one might add that some communications systems utilize external modulators, and the laser is used only as a CW radiation source. The advantages of such a scheme are (1) modulation frequency is not limited by electronic lifetimes, and (2) small temperature variations in the laser caused by changing modulation currents can be eliminated. Such small

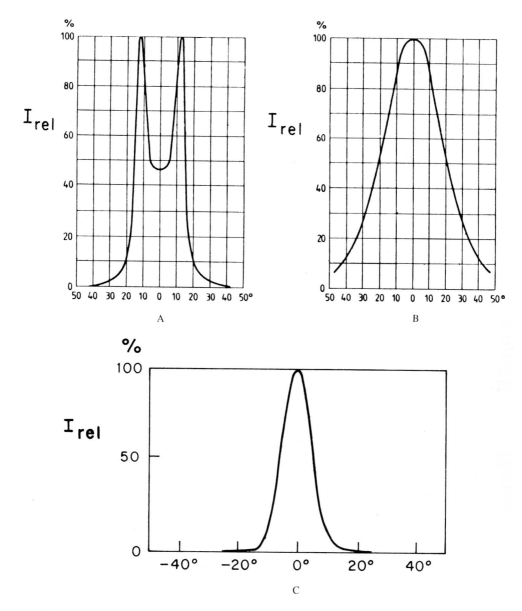

FIGURE 14. Far-field distributions of laser diodes; (A) gain-guided laser parallel to junction plane; (B) gain-guided laser perpendicular to junction plane; (C) index-guided laser parallel to junction plane; (D) index-guided laser perpendicular to junction plane.[16,17]

changes in temperature shift the emitted wavelength, producing undesirable effects in frequency and phase modulation systems.

E. Additional Optical Characteristics

For the sake of completeness, one should briefly mention the polarization and coherence properties of optical sources since these play an important role in some modulation schemes.

Surface emitting IREDs give by and large nonpolarized light. Because of the relatively broad bandwidth one can not talk about strict coherence either.

In contrast to this, laser diodes emit a linearly polarized TE-mode with the electric vector parallel to the p-n junction plane. The degree of polarization may vary somewhat over the entire far-field. One also has to keep in mind that some spontaneous emission light is

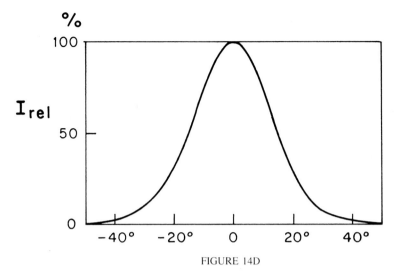

FIGURE 14D

indispensable to the operation of any laser, so every laser output will contain a small fraction of nonpolarized light. This small fraction is somewhat higher for gain-guided, than for index-guided structures.

Coherence properties also depend on the nature of lateral guidance and on whether or not the laser oscillates in a single or several modes. Coherence lengths for single-mode lasers can extend from several meters to several kilometers, while this distance shrinks to a few centimeters for multimode lasers. In the latter case, the widths of an individual mode and the number of modes determine the coherence length. Since the width of a mode decreases with the increase of optical power, the coherence improves with pumping.

F. Noise

The generation of photons in IREDs and laser diodes is based on a quantum statistical process: the recombination of carriers which by its very nature is subject to fluctuations. Additional noise is created by small fluctuations in the pumping current and variations in temperature. Electroluminescent diodes that utilize only spontaneous recombination and also emit fewer photons per unit solid angle, are generally much noisier than laser diodes. Assuming 100 μW power coupled into a fiber at about 0.9 μm and assuming a Poisson noise process, the purely quantum statistical noise leads to a relative noise intensity of about -70 dB for the entire bandwidth. An actual situation can be worse due to the other noise processes mentioned above.

The noise performance of laser diodes is considerably better than IREDs. First, most of the output is generated by stimulated recombination which in the theoretical limit is noise free. Only the always present spontaneous emission light accounts for intensity fluctuations. Second, more light can be coupled into the fiber due to the larger output and narrower far-field of the laser.

Noise frequency spectra of a laser follow the resonant characteristics of the modulation spectrum of the device (Figure 17). An interesting feature of the noise behavior of laser diodes is a sharp increase in noise output near threshold.[20] After the threshold has been passed the noise falls back rapidly. The overall relative noise intensity depends again on the nature of the lateral guidance.[21] In index-guided structures, the relative noise level per unit bandwidth approaches the theoretical limits of quantum statistical noise and may have values as low as 10^{-16} to 10^{-17}.[21] The lateral gain profile of the gain-guided laser, on the other hand, and the shape of its optical resonator is determined by the carrier distribution, which in turn depends on current flow and diffusion. These additional random processes and the multimode nature of the emission produce excess noise that brings the total relative

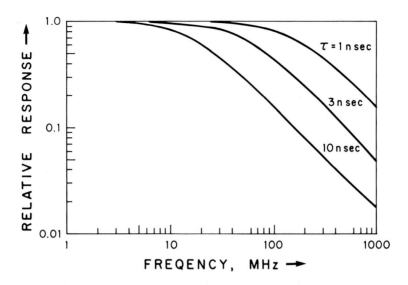

FIGURE 15. Light output response of an IRED. τ is the recombination lifetime.

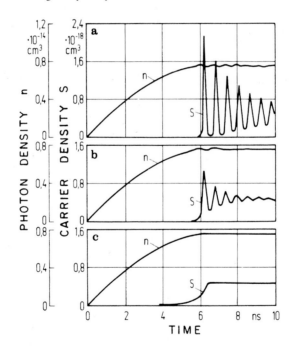

FIGURE 16. Response of laser diode to step current input. n: carrier density; S: photon flux. (a) $\beta = 10^{-4}$; (b) $\beta = 10^{-3}$; (c) $\beta = 10^{-2}$. β: parameter characterizing the optical cavity. (After Winstel, G. and Weyrich, C., *Optoelektronik I*, Springer-Verlag, Basel, 1980. With permission.)

noise level per unit bandwidth to about 10^{-14}.[21,22] This figure refers to the noise measured in all modes. Referred to a single mode, the noise is a thousand times higher.[22]

Besides the intensity fluctuations, one has to deal with small variations in the emitted wavelength. This noise is particularly objectionable in PCM and phase-modulated systems. It is produced mainly by variations in temperature that changes the bandgap and the optical constants of the material. Wavelength fluctuation can appear either as a random shifting of

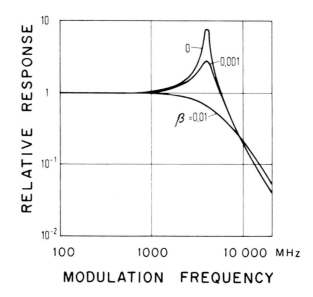

FIGURE 17. Frequency response of a laser diode. β as in Figure 16. (After Winstel, G. and Weyrich, C., *Optoelektronik I*, Springer-Verlag, Basel, 1980. With permission.)

the entire spectrum and/or as changes in the subdivision of the total optical output into individual longitudinal modes (mode competition noise). In single mode lasers, this mode competition noise causes the mode to jump among several allowed longitudinal mode positions that results in an effective broadening of the spectrum.

G. External Factors Affecting the Operation of Sources

Among the external factors, temperature and optical feedback will be discussed. Since these effects are not too significant in electroluminescent diodes, the discussion will be restricted to laser diodes.

Temperature effects — Temperature affects practically all material parameters of a semiconductor that are important for the operation of a laser. First, the bandgap decreases with increasing temperature that has a major effect on the imaginary part of the complex refractive index (gain and/or absorption). The real part (refractive index) changes also, affecting the optical parameters of the waveguide. The change in bandgap alone causes a shift in wavelength of a few angstroms per degree centigrade. In monomode lasers, this shift consists of a small gradual change in wavelength followed by a sudden jump as the resonator seeks out the particular longitudinal mode that has the highest gain under the given circumstances. In addition to the change in wavelength, the temperature also influences the output power through its effect on the threshold current. The threshold current is proportional to $\exp(T/T_o)$ with T_o an empirical constant.[23] T_o varies from 50K to 200K; the lower values are characteristic of InGaAsP (1.3 to 1.6 μm) lasers, the higher ones to GaAs or GaAlAs lasers. T_o is the most important parameter determining the thermal stability of the laser. As the fractional change in threshold current is inversely proportional to T_o, high values are desirable. The temperature rise also reduces the differential quantum efficiency, but this effect is far less drastic than the change in threshold. All these factors combine to give an overall decrease in output if the temperature rises. Some laser diode modules contain a small, monitor photodiode that can be used in a feedback loop to control the drive current if one wishes to maintain the average output of the laser at a constant level. Thermal runaway is still possible, however, if the value of T_o is not sufficiently large, since an increase in current produces more losses and ever increasing temperatures. To counteract this eventuality, the laser can be mounted on a small Peltier cooler that maintains a co[...] temperature. If T_o is large, say >150K, external cooling is not always necess[...] the application is particularly critical. The effect of temperature on excess noise was [...] discussed in the previous paragraph.

Optical feedback — Laser diodes employed in fiber optic communication systems are subjected to optical feedback, that is the back reflection of their own light coming from various optical discontinuities in the transmission system. Such discontinuities exist at the coupling point between laser and fiber, laser and focusing optics, couplers between fibers, etc. Optical feedback deteriorates almost all operating characteristics of the laser, changes the output power and the linearity of the light-current characteristics and increases noise. The severity of these effects depends on the amount and nature of feedback light and on the type of the laser used. Single-mode lasers with rigid, index-guided resonators tend to react much more drastically to feedback than multimode lasers; they can become very noisy and switch into a multimode oscillatory mode. In fact, as a consequence of optical feedback, they can lose many of their advantages over multimode lasers. It is possible to eliminate feedback in monomode systems by the use of an optical isolator between laser and transmission system. Isolators make use of the polarized nature of laser light and rotate the plane of polarization of the feedback light by $\pi/2$ which can then be filtered out by a polarizer. Multimode, gain-guided lasers are far less sensitive to feedback and may not require the use of optical isolators.

IV. PHYSICAL PRINCIPLES AND STRUCTURE OF ELECTROLUMINESCENT DIODES

The simplest form of an electroluminescent diode is a forward-biased p-n junction. Electrons diffusing over the barrier recombine with holes in the p-region. Such a diode emits light all along the junction region and also over its top surface if the junction is not located too deep in the material and if the contact allows the radiation to escape. The external quantum efficiency of the device is proportional to the number of photons emitted by the diode in unit time divided by the number of carriers transported through the device in unit time,

$$\eta_{ext} = eN_{ph}/I \tag{16}$$

The number of photons N_{ph} can be determined from the total emitted power of the diode as

$$N_{ph} = P/\overline{h\nu} \tag{17}$$

where $\overline{h\nu}$ is the average photon energy for the luminescence band. Three loss mechanisms are involved in reducing the external quantum efficiency: (1) the intrinsic quantum efficiency of the material, in other words, the fact that not all recombination processes are radiative; (2) not all carriers injected into the diode participate in recombination; and (3) not all photons generated in the device can be emitted.

Some nonradiative effects were already mentioned in Section II. Among the current related loss mechanisms one can mention the injection efficiency of the p-n junction, recombination current in the barrier region of the junction that is usually nonradiative, and leakage currents on the surface. The optical losses are probably the most significant in a simple IRED. Since spontaneous recombinations are random, the directions of the generated photons are also perfectly random. Some do not even travel toward an emitting surface and become absorbed in the substrate or other structural elements. Because of the high refractive index of GaAs (3.6) or GaAlAs (3.5), photons which travel at a larger angle than the critical ($\theta_c = \sin^{-1}$

FIGURE .   .etries to improve optical emission efficiency of luminescent diodes.[24] (a) Flat surface emitter; (b) hem.. .erical radiator; (c) emitter with Weierstrass sphere ($r_2 = r_1/\bar{n}$); (d) ellipsoid-radiator. (After Winstel, G. and Weyrich, C., *Optoelektronik I*, Springer-Verlag, Basel, 1980. With permission.)

$(1/\bar{n}) = 16°)$ from the normal to the emitting surface become totally reflected. This effect alone reduces the number of emittable photons to about 2% of the total emanating from an isotropic, point-like source located somewhere inside the diode. The transmission coefficient on the surface and absorption in the material further reduce the emission. An appropriate shaping of the emitting surface (Figure 18) or attaching suitably focused microlenses to the device improve the optical efficiency and the coupling of the emission into a fiber.

The use of a heterojunction and multilayered structure can greatly improve the efficiency of IREDs. Naturally, this improvement is bought at the cost of a much more elaborate manufacturing process. Such a diode contains a p- or n-doped GaAs active layer sandwiched between a p-$Ga_{0.7}Al_{0.3}As$ and an n-$Ga_{0.7}Al_{0.3}As$ layer. The GaAs may sometimes contain a few percent aluminum to shift the emission wavelength to a smaller value. This double-heterostructure (DH structure) improves several operating parameters of the device simultaneously: (1) the p-n heterobarrier greatly increases the electron injection efficiency; (2) injected carriers become confined between the two heterojunctions and cannot diffuse into the device that makes the whole recombination process more efficient and more controllable; and (3) the generated photons must pass through the GaAlAs on their way to the emitting surface. Since GaAlAs has a somewhat larger bandgap than the active layer, no absorption can take place there. The multilayer structure is grown on a GaAs substrate. Usually the device is mounted "upside down" with the heat generating layers closer to the heat sink. Light can be coupled out through a hole etched into the GaAs substrate. The optical fiber is directly cemented into the hole providing greater coupling efficiency[25] (Figure 19). Double-heterostructure IREDs can also be built as edge-emitters.[26] Their emission characteristics then become similar to laser diodes and display a more directed far-field emission pattern.

V. PHYSICAL PRINCIPLES AND STRUCTURE OF DH LASER DIODES

Laser diodes operating in the CW mode at room temperature are almost exclusively fabricated as multilayer double heterostructures similar to the DH IREDs (Figure 19). Therefore, this discussion will be restricted to such DH structures. The operation of a laser diode below threshold is no different from a DH IRED. The principal differences appear only in the lasing region, above the threshold. Laser diodes utilize stimulated recombination processes as well as spontaneous emission; therefore, current and carrier densities must be much higher in these devices. In order to produce coherent radiation, lasers require an optical resonator in which the EM wave is effectively confined and guided and in which a well defined standing wave pattern can develop. These requirements reflect then on the design and dimensions of laser diodes that are somewhat different from IREDs. The optical cavity of these lasers is of the Fabry-Perot type that is formed by two (1,1,0) cleavage planes of

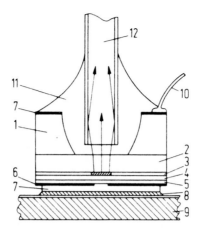

FIGURE 19. Surface-emitting DH IRED (Burrus type). (1) n-GaAs substrate; (2) n-$Ga_{0.7}$ $Al_{0.3}$ As; (3) p-$Ga_{0.95}$ $Al_{0.05}$ As; (4) p-$Ga_{0.7}$ $Al_{0.3}$ As; (5) p-GaAs; (6) SiO_2 isolation layer; (7) gold contact; (8) In solder; (9) gold plated ceramic; (10) gold lead; (11) epoxy cement; (12) glass fiber. (After Winstel, G. and Weyrich, C., *Optoelektronik I*, Springer-Verlag, Basel, 1980. With permission.)

the crystal. These two planes represent semitransparent mirrors which due to the high refractive index of the semiconductor reflect about 30% of the light power back into the optical resonator to sustain the stimulated emission process. This is in stark contrast to gas lasers that feed about 90% of the power back into the cavity and emit only about 10%. The ratio of feedback to output power can be altered by coating the end mirrors with a dielectric layer of suitably chosen thickness, but such changes also bring some penalties with them. Reducing the transmission of the end mirror reduces threshold current requirements because of the increased optical feedback. One loses precious output power, however, with its predictable undesirable consequences on signal-to-noise ratio. Increasing transmission has the opposite effect, and increased threshold current requirements deteriorate the heat balance of the device. Although dielectric coating of mirrors is frequently employed to protect the mirrors against damage and quick deterioration, the 0.3/0.7 feedback-output ratio is seldom drastically changed.

The typical dimensions of a laser diode cavity are the following: in the direction of wave propagation (longitudinal direction) 200 to 500 μm, in the direction perpendicular to the junction plane (transverse direction) 0.1 to 0.2 μm, in the direction parallel to the junction plane (lateral direction) 3 to 12 μm. The partial confinement of the EM wave within these boundaries gives rise to well defined field distributions that are called cavity modes. According to the boundary that produces such a mode, one speaks about a longitudinal, transverse, or lateral mode. Because of the large differences between the longitudinal, lateral, and transverse dimensions, it is possible to uncouple the three-dimensional wave guiding problem and treat each direction separately in an approximate calculation. This is particularly permissible if one is mainly interested in conditions prevailing near the optical axis.

A. Waveguiding in Laser Diodes

In the longitudinal direction, the modes are the well-known Fabry-Perot resonator modes. The condition for obtaining a standing wave pattern is

$$\ell = i\, \lambda_o/(2\bar{n}) \tag{18}$$

with ℓ: cavity length, λ_o: vacuum wavelength, and i an integer. The spacing between these modes, considering also the wavelength dependence of $\overline{n}$ is given by

$$\Delta\lambda_o = \lambda_o^2 \left[2\ell \left(\overline{n} - \lambda_o \frac{d\overline{n}}{d\lambda} \right) \right] \tag{19}$$

This spacing varies between 1.5 to 2.5 Å for GaAs or GaAlAs lasers depending on the cavity length ℓ. For 1.3 to 1.6 μm InGaAsP/InP lasers, the spacing is considerably larger. Which one or how many of these possible longitudinal modes will actually lase depends on the spectral distribution of the gain (Figure 7). The bandwidth of the gain is several hundred angstroms wide so, at least in principle, hundreds of longitudinal modes are conceivable. Their actual number is much less than that, however, because the gain has to be sufficiently large to overcome the optical losses of the cavity, such as residual absorption, scattering from defects and irregularities and last, but not least, the emitted power which represents an optical loss from the resonator's point of view. These conditions narrow down the spectral width of the available gain considerably; furthermore, threshold current conditions also vary somewhat with wavelength which affects the selection of modes. The longitudinal mode selection is still not a fully understood process in semiconductor laser diodes; it appears that the stability of the lateral guidance plays an important role: lasers having a rigid lateral guidance, such as the index-guided types, prefer to select one longitudinal mode, while lasers with "soft" lateral guiding mechanisms, such as provided by carriers, tend to oscillate with 20 or 30 longitudinal modes simultaneously.

The waveguiding in the transverse direction is assured by the finite step in the refractive index between the active region (GaAs or low Al concentration GaAlAs) and the two cladding GaAlAs layers with a higher Al content. (Figure 20). The bandgap increases and the refractive index decreases with aluminum in the GaAs. The solution of the wave equation for such a slab waveguide can be found in several texts.[2,3] A strongly abbreviated treatment employing symbols defined in Figure 20 will be followed here.

The solution of the wave equation in the active region leads to cosine (even modes) and sine (odd modes) functions. For the transverse electric (TE) polarization, the most frequently encountered in these diodes, one obtains

$$\mathcal{E}_y(x) = A_e \cos(\kappa x) \exp(j\omega t - \beta_z z) \tag{20}$$

$$\kappa^2 = \overline{n}_2^2 (2\pi/\lambda_o)^2 - \beta_z^2 \tag{21}$$

$$\mathcal{E}_y(x) = A_o \sin(\kappa x) \exp(j\omega t - \beta_z z) \tag{22}$$

β_z is the propagation constant in the z-direction and it will be found to be somewhat less than $2\pi \, \tilde{n}_2/\lambda_o$ as the modes travel slightly off-axis. If one requires confining the wave to the vicinity of the middle layer, the solution in the cladding layer must necessarily decay exponentially:

$$\mathcal{E}_y(x) = B_{e,o} \exp[-\gamma(|x| - |d/2|)] \exp(j\omega t - \beta_z z) \tag{23}$$

with

$$\gamma^2 = \beta_z^2 - \overline{n}_1^2 (2\pi/\lambda_o)^2 \tag{24}$$

Satisfying dielectric boundary conditions at the interface leads to transcendental eigenvalue equations for the even and odd modes:

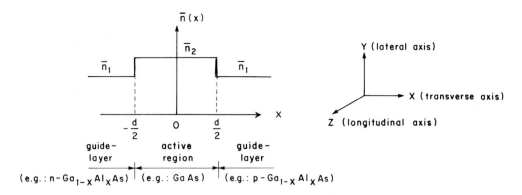

FIGURE 20. Symmetrical slab waveguide.

$$\tan(\kappa d/2) = \gamma/\kappa$$

and

$$\tan(\kappa d/2) = -\kappa/\gamma \tag{25}$$

Because of Equations 21 and 24

$$(\bar{n}_2^2 - \bar{n}_1^2)(2\pi/\lambda_o)^2 = \kappa^2 + \gamma^2 \tag{26}$$

Introducing $\kappa d/2$ and $\gamma d/2$ as dimensionless variables, Equations 25 and 26 lead to a simple graphical solution shown in Figure 21. The curves represent Equations 25 and the radius of the circle is given by

$$[(\bar{n}_2^2 - \bar{n}_1^2)]^{1/2}\ (2\pi/\lambda_o)(d/2) \tag{27}$$

The number of possible solutions and the number of transverse modes increase with the slab thickness d and the index difference. To guarantee operation of the laser in the fundamental transverse mode, the index difference and slab thickness must necessarily be small. Intensity distributions corresponding to the five lowest order modes are shown in Figure 22.

The requirement to maintain fundamental mode operation brought about the extremely small dimensions (0.1 to 0.2 μm) for the thickness of the slab. This also reduced current requirements, a very desirable effect; still the production of such thin layers with the required homogeneity, uniformity in thickness and smooth, high quality interfaces was no minor technological achievement.

The thickness of the slab and the step in the refractive index also influences other laser characteristics. Since lasers utilize stimulated recombination, effective interaction between carrier and photons can take place only when the injected carriers and the optical field are restricted more or less to the same volume. The carriers in a DH structure become confined between the two heterobarriers, that, is in the middle layer of the slab. The EM wave, on the other hand, leaks out with its exponential tail into the cladding layers that are not pumped. The effect of index step and slab thickness on the shape of the fundamental mode is shown in Figure 23. The effectiveness of coupling between carriers and photon field is expressed by the confinement factor

$$\Gamma = \frac{\displaystyle\int_{-d/2}^{d/2} |\mathscr{E}_y|^2 dx}{\displaystyle\int_{-\infty}^{\infty} |\mathscr{E}_y|^2 dx} = \left\{ 1 + \frac{\cos^2(\kappa d/2)}{\gamma\left[\dfrac{d}{2} + \dfrac{1}{\kappa}\sin(\kappa d/2)\cos(\kappa d/2)\right]} \right\}^{-1} \tag{28}$$

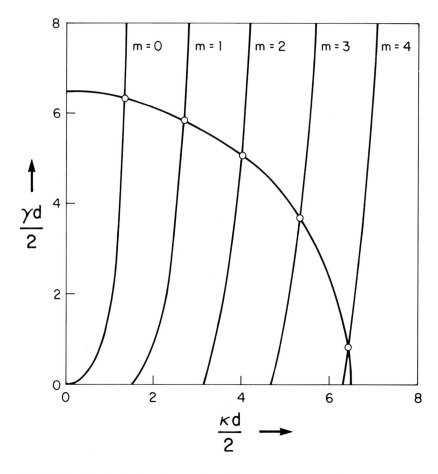

FIGURE 21.   Solution of eigenvalue equation of slab waveguide; m: order of mode; intersections are solutions of Equations 25 and 26.

There is a much larger variety of methods and solutions for lateral confinement than for transverse confinement. The technologically simplest solution is the stripe geometry laser. In its various forms, the current feed on the p-side of the diode is restricted to a narrow (3 to 12 μm) wide strip. Some examples for these structures are shown in Figure 24. The sequence of layers is similar to those already shown for the DH IRED. Layer 3 is the active region that may also contain some aluminum to shift the emission to shorter wavelengths. Layer 5 is used mainly as a contacting layer, since it is easier to make good contacts to p-GaAs than to p-GaAlAs. Layer 4 is made thick enough (1.5 to 2 μm) to avoid penetration of the exponential tail of the fundamental transverse mode into the GaAs where absorption losses would be too high. The structures shown provide means to confine current flow to the central regions of the device: in (a) this is accomplished by an etched mesa; in (c) the conductivity of photon bombarded regions is destroyed while the conductivity of the central region is protected by a wire-mesh during bombardment; in (d) the p$^+$−n junction in layer 2 has a higher barrier than the junction between layers 3 and 2 because of the high p$^+$ doping on the sides; therefore, the current will flow mainly through the center; in (e) the reverse biased p-n junction between layers 4 and 6 is broken through only in the center by the Zn-diffusion. In (b) which is the simplest structure, lateral current spreading is not prevented by any physical means other than the stripe contact, but it can be held within tolerable limits by careful design. There is no lateral optical waveguide built into these

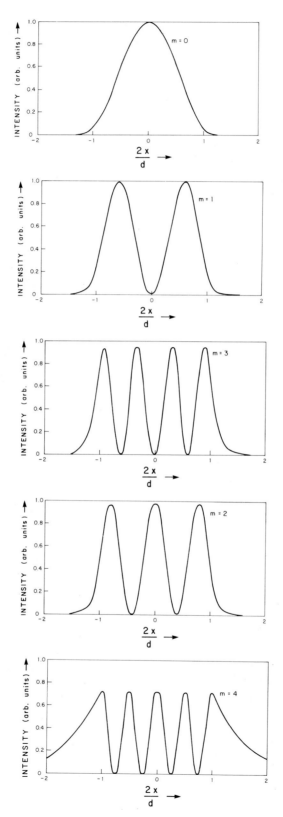

FIGURE 22. Normalized intensity distributions of a symmetrical slab waveguide. Modes correspond to solutions shown on Figure 21.

lasers; the waveguide is established by the pumping current. Since the current density and the resulting carrier density will have their maxima in the center, the resulting optical gain profile, which is a simple function of the carrier density, will also peak on the optical axis and fall off towards the sides. This gain profile is the essential lateral guiding mechanism in these lasers and hence the name: gain-guided laser. The solution of the EM wave equation for these cases is no elementary task and the reader is referred to the extensive literature.[27-29] In addition to a convex gain profile, the real part of the refractive index has a slight local minimum on the optical axis that tends to counteract the positive guiding provided by the gain ("antiguiding"). The complicated optical index profile is responsible for nonplanar phase fronts in the guide,[30] for the higher astigmatism of the beam, and for the peculiar far-field emission pattern of these lasers. Fundamental lateral mode operation and stable light-current characteristics can be attained only with very narrow (<8 μm) stripes. Wide stripes allow the development of higher order lateral modes (Figure 25) or lead to filamentation that results in kinks or instabilities in the light-current characteristics.

The multimode spectrum, higher astigmatism, and wider far-field pattern are distinct disadvantages of the gain-guided laser. Its theory is also more complicated, but its manufacturing technology, particularly that of the oxide-stripe laser (Figure 24b) is the simplest, involving the least complications and critical steps in the manufacturing process. This is a distinct advantage if one wishes to produce lasers having high reliability, stability, and low aging. They are also the least sensitive to optical feedback.

The desire to provide better lateral guidance for the EM wave and to obtain single-mode operation leads to the development of the so-called index-guided structures. A positive step in the real part of the complex refractive index is built into the lateral structure of the device, thus the wave experiences the same kind of guiding in the lateral direction as in the transverse direction discussed earlier. Typical examples for these are the buried heterostructures (BH) shown in Figure 26. In (a) a curved groove is provided in the substrate which produces — by the proper control of the growth process — a pinched off active region so that it is completely surrounded by GaAlAs that has a lower refractive index.[32] In (b) which is a more frequently encountered BH structure, a narrow mesa is etched into the usual four-layer sequence, after which the device is returned to the growth furnace and p-GaAlAs is regrown in the etched away areas.[33] In (c)[34] a similar process is employed, except the mesa etch follows after the growth of the active region. This latter structure has two additional features worth mentioning. The stripe contact is laterally bounded by a reverse biased p-n junction that eliminates the need for an isolating oxide layer. The n-GaAlAs cladding layer consists of two parts and forms, with the other layers, a so-called large optical cavity (LOC). The upper n-GaAlAs layer can be extended beyond the laser mirror and can be used to couple the radiation into another integrated optical device. For conventional laser operation, this is not needed.

Waveguiding in these lasers, particularly in (b) and (c), can be simply calculated on the basis of the slab waveguide model. Because of the relatively large step in the refractive index, confinement is strong and fundamental lateral mode operation can be assured only with very narrow stripes (1 or 2 μm). This reduces current requirements, a welcome feature, but also limits the output power. The optical output power density must be limited in any semiconductor diode to avoid catastrophic mirror damage. BH-lasers are characterized by single longitudinal mode operation, very low threshold currents, high differential quantum efficiency, highly linear output characteristics, and low astigmatism. They are limited in their output power, however; their thermal stability is somewhat inferior since they cannot be mounted "upside down" on the heat sink, and they are extremely sensitive to optical feedback because of the high Q of their optical resonators.

An interesting form of lateral waveguiding and a highly successful laser structure is

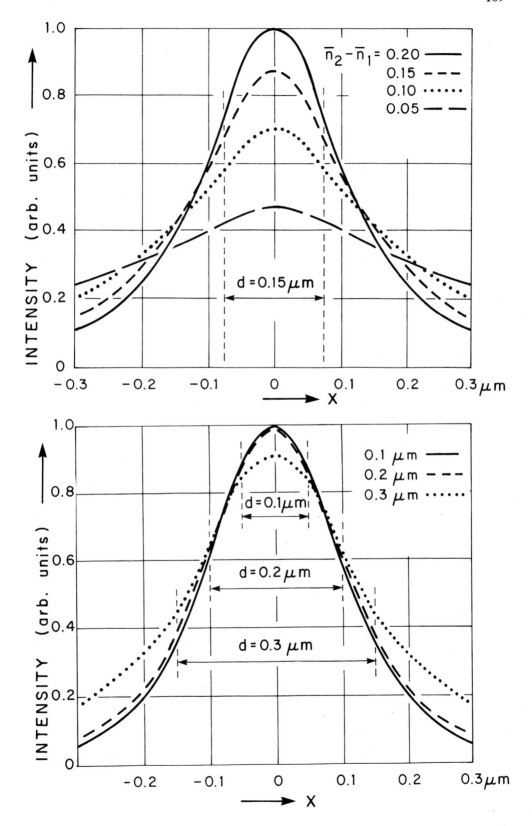

FIGURE 23. Effect of refractive index step ($\bar{n}_2 - \bar{n}_1$) and slab thickness d on the shape of the normalized fundamental mode.

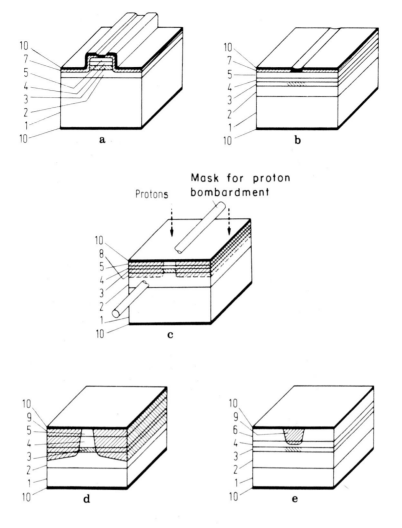

FIGURE 24. Stripe geometry laser diodes. (1) GaAs substrate; (2) n-GaAlAs; (3) GaAs active region may be p or n doped; (4) p-GaAlAs; (5) p$^+$GaAs; (6) n-GaAs; (7) isolating oxide layer; (8) proton bombarded layers; (9) Zn-diffused layers; (10) metallic contacts. (After Winstel, G. and Weyrich, C., *Optoelektronik I*, Springer-Verlag, Basel, 1980. With permission.)

illustrated in Figure 27.[35] In this channeled substrate planar (CSP) laser a narrow (5 μm) groove is etched into the substrate. As a result of this, the n-GaAlAs cladding layer is thinner outside the channel so as to allow leakage of the transverse mode into the lossy GaAs substrate. Thus, the losses of the waveguide are higher outside the channel region. In the channel, the waveguide corresponds to that of a regular stripe laser. Such a guide can be called a gain/loss guide. Because the effective refractive index of the guide $\bar{n}_{eff} = \lambda_o\beta_z/(2\pi)$ as defined by the propagation constant β_z, is higher in the central channel region than outside it, this structure can also be referred to as "index-guided". Regardless of nomenclature, the main difference between the CSP and the regular gain-guided laser is in the way in which the lateral profile of the optical constants is realized; in the ordinary gain-guided laser, this profile is created by current-pumping, and thus it is subjected to statistical fluctuations produced by temperature variations and other stochastic processes. In the CSP-laser, the lateral profile of the optical constant is firmly embedded in the device by a grooved channel. The difficulty of producing this structure is in controlling the depth of the groove

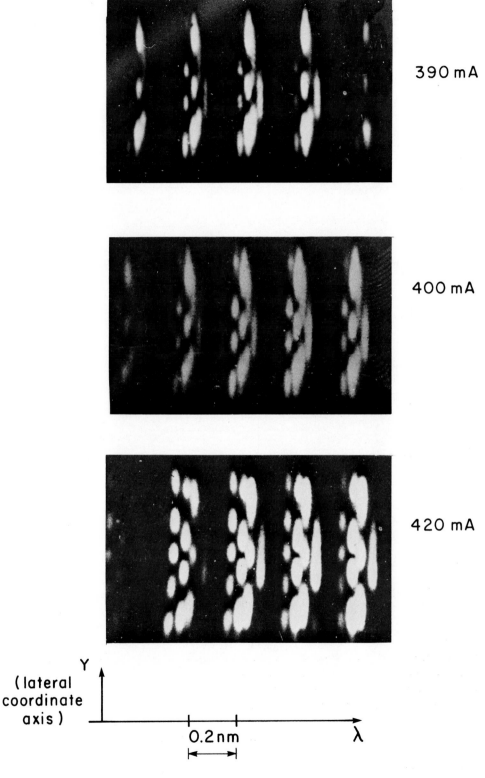

390 mA

400 mA

420 mA

Y

(lateral
coordinate
axis)

0.2 nm

λ

FIGURE 25. Development of higher order lateral modes with pumping current in a wide mesa-type laser.[31] The photos show laterally and spectrally resolved near-field patterns.

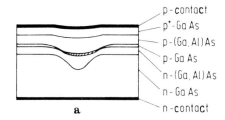

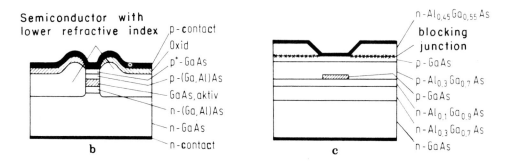

FIGURE 26. Buried heterostructure (BH) lasers. (After Winstel, G. and Weyrich, C., *Optoelektronik I*, Springer-Verlag, Basel, 1980. With permission.)

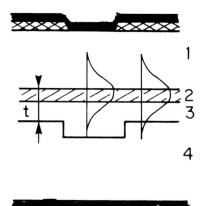

FIGURE 27. Channeled substrate planar (CSP) laser.[35] (1) $pGa_{1-x}A\ell_xAs$; (2) active layer; (3) $n\,Ga_{1-x}A\ell_xAs$; (4) substrate. (After Winstel, G. and Weyrich, C., *Optoelektronik I.*, Springer-Verlag, Basel, 1980. With permission.)

and maintaining the required thickness t that has rather stringent tolerances. Furthermore, the mask for contacting has to be indexed exactly over the channel after it has been covered by four epitaxial layers. There are several varieties of CSP-structures, each of which is an attempt to simplify one or the other critical step in the manufacturing process. The principle of their operation is not significantly different from the basic CSP type.

The operating characteristics of CSP lasers are similar to those of the BH-lasers. They oscillate in a single longitudinal mode, have a high differential quantum efficiency, highly linear output characteristic, and low astigmatism. Their output power is higher than that of the BH laser, because of the wider optical resonator, but this also somewhat increases the threshold current requirements. They are also very sensitive to optical feedback.

B. Threshold Current and Lasing

Optical absorption is turned into gain in an optically active medium when the rate of downward transitions from the higher energy level to the lower energy level exceeds that of the upward transitions. It can be shown that in the case of a semiconductor this requirement postulates that the difference between the quasi-Fermi levels of electrons and holes must be equal to or larger than the energy of the emitted photons, which in turn corresponds to the energy gap of the semiconductor. In order to satisfy these conditions one or both of the injecting p- and n-layers must be degenerate and the forward bias voltage applied to the diode must exceed the energy gap. For GaAs or GaAlAs devices this gives a forward voltage of about 1.5 V. Series resistance effects increase the voltage somewhat. These high forward voltages mean that injection laser diodes operate at very high current densities. Early models, before the DH-structure was invented required several times 10^4 A/cm² to reach lasing and could not be operated CW at room temperature. The development of the DH structure[36-38] reduced this value to about 10^3 A/cm² or less. Figure 28 shows the energy band diagram of a p-GaAlAs/p-GaAs/n-GaAlAs DH structure without and with forward bias. The hetero-barriers ΔE_1 and ΔE_2 prevent the diffusion of the carriers from the recombination region; thus, all the injected electrons can be fully utilized.

The appearance of a positive gain in the active part of the semiconductor is not sufficient to produce lasing. The gain has to be large enough to compensate for all the losses encountered by the optical wave in the resonator. Among these losses are residual absorption in the medium, mainly free carrier absorption, scattering losses due to defects and geometrical irregularities of the waveguide, and last but not least, the output photon flux leaving the resonator through the end mirrors. The condition at the lasing threshold is that the gain encountered by a photon during a round trip in the cavity exceeds the losses:

$$R^2 \exp(2\ell g - 2\ell\alpha_i) \geq 1 \qquad (29)$$

which leads to

$$g \geq \alpha_i + \frac{1}{\ell} \ell n\left(\frac{1}{R}\right) \qquad (30)$$

with α_i due to the internal optical losses of the cavity, mainly free carrier losses, R the reflection coefficient of the end mirrors, and ℓ the cavity length. The last term in Equation 30 is often denoted by α_m, the mirror loss coefficient, which of course represents the useful output (per unit cavity length) emitted through the two end mirrors. If one wishes to relate the required gain to the pumping current density, one has to realize that optical losses are encountered by the entire cross section of the EM wave while pumping is confined to the active (recombination) region of the cavity. This can easily be accounted for with the confinement factor (Equation 28). Furthermore, the gain and some of the optical losses depend on the intensity of the optical wave as well as the carrier density that are not constant in the lateral direction in some stripe geometry lasers. It is assumed, therefore, that g and α_i correspond to a properly weighted average taken over the lateral direction.

The relationship between the gain coefficient g and carrier density n based on Reference 10 can be written as [39]

$$g = an^2 - b \qquad (31)$$

where a and b are wavelength dependent constants. Theoretically, a and b can be evaluated or determined from the curves of Figure 7. In the steady state, the pumping current density is related to the carrier density if one assumes both bimolecular and linear recombination mechanisms as

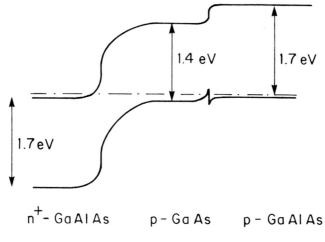

n⁺- GaAlAs   p - GaAs   p - GaAlAs

A

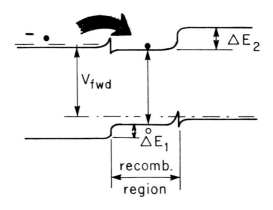

B

FIGURE 28.   n-GaAlAs/p-GaAs/p-GaAlAs heterobarrier; (A) in equilibrium; (B) under large forward bias. ΔE_1, ΔE_2 barriers for holes and
electrons, respectively.

$$j = ed(Bn^2 + n/\tau) \qquad (32)$$

with e the electronic charge, d the thickness of the active layer, and B and τ as defined in
Section II. A. Combining Equations 29, 30, and 31, one obtains

$$j_{th} = ed\left\{\frac{B}{\Gamma a}\left[\alpha_i + \frac{1}{\ell}\ell n\left(\frac{1}{R}\right)\right] + \frac{Bb}{a} + \sqrt{\frac{1}{\Gamma a\tau^2}\left[\alpha_i + \frac{1}{\ell}\ell n\left(\frac{1}{R}\right)\right] + \frac{b}{a\tau^2}}\right\} \qquad (33)$$

If bimolecular recombination dominates, as is the case with lightly doped active layers,
which is the customary doping in up-to-date lasers, the first two terms of Equation 33 will
dominate. The threshold current increases linearly with d and this also explains why extremely
narrow (about 0.1 to 0.2 μm) active layers are currently used. Further reduction of d does
not reduce the threshold current further, because the EM wave begins to leak out into the
cladding layers (see Figure 23), thereby rapidly reducing Γ. An optimum lies in the vicinity
of 0.1 μm.

The slope of the light-current characteristics in the lasing region, the differential quantum efficiency, can be calculated as

$$\eta_d = \frac{dP_L}{dI} \tag{34}$$

The output optical power P_L is usually referred to the output on one side of the laser, as only that is used in practical applications. The other side is often employed for monitoring or controlling purposes. The differential quantum efficiency η_d is proportional to the energy of an emitted photon, the internal quantum efficiency of the stimulated recombination process η_{st}, and the probability of such a photon being emitted

$$\eta_d = \left(\frac{h\nu}{2e}\right) \eta_{st} \frac{\alpha_m}{\alpha_i + \alpha_m} = \left(\frac{h\nu}{2e}\right) \eta_{st} \left[1 + \frac{\alpha_i \ell}{\ell n(1/R)}\right]^{-1} \tag{35}$$

The factor 1/2 takes the one-sided emission into account. For a typical case with $\alpha_i = 40$ cm^{-1}, $\ell = 400$ μm, R = 0.3, one obtains $\eta_d = 0.3$ W/A when $\eta_{st} \approx 1$.

The operation of the laser above threshold can be best described by rate equations that are really conservation equations and represent the balance between carriers and photons

$$\frac{\partial n}{\partial t} = -R_{sp}(n) - \sum_i g_i(n)S_i + j/(ed) \tag{36}$$

$$\frac{\bar{n}}{c} \frac{dS_i}{dt} = \beta_i R_{sp}(n) + g_i(n)S_i - S_i \frac{\bar{n}}{\tau_p c} \tag{37}$$

The first equation expresses the conservation of carriers. The first term on the right hand side $R_{sp}(n)$ is the carrier loss due to spontaneous recombination, the second the loss due to stimulated recombination with $g_i(n)$ being the gain coefficient and S_i the photon flux of the i-th mode (in m^{-2}sec^{-1}), and j/ed is the rate at which carriers are injected by the pumping current density. Equation 37 gives the balance for photons in each of the i lasing modes, τ_p is the photon lifetime and thus represents photon loss processes. The value of τ_p is of the order of picoseconds for typical semiconductor lasers, about a thousand times less than electron lifetimes. The fraction of the spontaneous recombinations that happen to fit the i-th mode is β_i. It is mainly a function of resonator geometry and plays an important role in the dynamic response of the device (see Figure 16). A photon balance equation of the type of Equation 37 has to be written for each longitudinal mode. It is assumed that all quantities in these equations are averaged over their respective spatial variation. The solution of these equations is a rather formidable task even for a single-mode laser (i = 1). Numerical solutions can be obtained on a computer provided that all functional relationships and all parameters are known. A characteristic feature of the solutions in the steady state (d/dt = $\partial/\partial t = 0$) is that the carrier density and, therefore, the gain rises with the pumping current until threshold is reached when it saturates (gain saturation) (Figure 29). The normalized output photon flux as a function of normalized pumping current on a double-log scale is shown in Figure 30. On a linear scale, this plot gives the familiar light-current characteristic of the laser. The effect of the cavity parameter β is also visible. Higher values of β make the transition between luminescence and laser operation less abrupt, the changeover more rounded. Solving the rate equations for a small sinusoidal driving current gives the small-signal frequency response discussed in Section III.

C. Aging Characteristics

All light emitting devices display aging phenomena that appear mainly as a reduction of

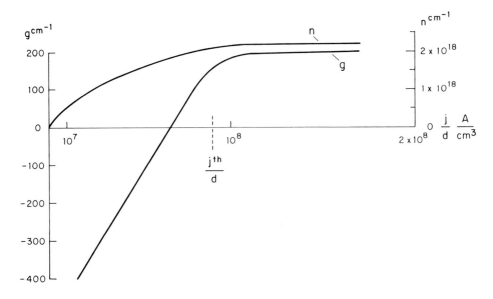

FIGURE 29. Saturation of carrier density and gain with injection above threshold. (Curves calculated with typical laser parameters.)

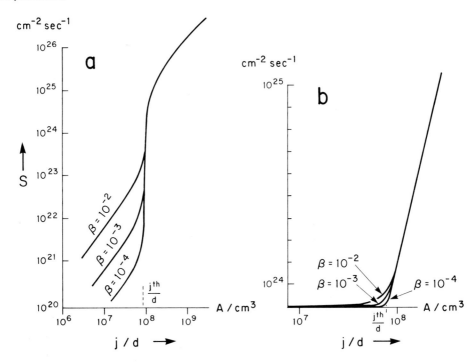

FIGURE 30. Normalized light-current characteristics on (a) double-log and (b) linear scales. β: cavity parameter. (Curves calculated with typical laser parameters.)

light output with time if the pumping current is held constant. Luminescent diodes are less plagued by aging since they generally operate at lower current and photon-density levels. A typical aging curve of a luminescent diode is shown in Figure 31. Useful life is considered to have expired when the light output has dropped to 50% of its initial value.

Aging is a much more serious problem in laser diodes, because the electrical and optical loads — current densities in the 10^3 to 10^4 A/cm^2 range, photon fluxes approaching 10^7

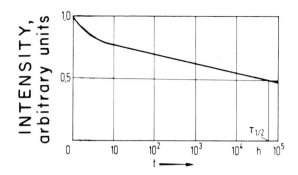

FIGURE 31. Reduction of light emission of a luminescent diode with time. (After Winstel, G. and Weyrich, C., *Opto-elektronik I*, Springer-Verlag, Basel, 1980. With permission.)

W/cm^2 — stress the materials to their limit of endurance. In fact, in the earlier stages of laser development, rapid aging appeared as the principal obstacle to practical applications. Lifetimes at room temperature were measured only in hundreds of hours. Degradation of the device is noticed again as a reduction of light output. This loss of power is due to two different effects in laser diodes: an increase of the threshold current and a decrease of the differential quantum efficiency. Of these two, the increase of the threshold is the more serious since a 10 to 15% change may push the device back into the nonlasing region of the characteristic unless the operating current is increased. This, of course, adds to the heating of the diode, and the process becomes even more accelerated. Thus the aging of a laser diode cannot be judged properly unless the light output is maintained at some constant level by a continuous readjustment of the operating current. Practical life expectancy can be better assessed by the increase with time of threshold current or the increase of the operating current needed to maintain a specified level of optical output.

Detailed investigations have shown that failure of laser diodes can be explained by three distinct mechanisms: (1) failure of the exit mirrors (facet degradation), (2) localized degradation of the optical resonator, or (3) slow, homogeneous degradation of the optical resonator. Failure of the mirrors is caused by the high optical flux densities and also by environmental effects. It is also known that facet degradation is enhanced by crystallographic defects and certain impurities on the surface. Coating the mirrors with an inert dielectric layer having a high dielectric strength, such as silicon- or aluminum-oxide can substantially improve conditions on the facets. Crystallographic defects may appear as a result of a faulty cleavage or improper growth of the epitaxial layers. As technology has improved significantly in the last 10 years many of these problems are under much better control today. The facet loading is mainly limited by the heating of the mirrors due to nonradiative surface recombination of carriers. Reducing this surface recombination could push the photon flux loading well above 10^7 W/cm^2. Improvements can most certainly be expected since the effort to obtain higher optical output is motivated not only by the optical communications industry that hopes to attain higher signal-to-noise ratios, but also by other applications such as laser data recording systems, fast printers, etc. that may become other important markets for laser diodes.

Localized degradation of laser diodes was perceived as the rapid development of dark spots in the electroluminescent image of the resonator. Since the GaAs substrate is partially transparent to the electroluminescent emission from the active region, the uniformity of the emission can easily be assessed with the aid of an infrared viewing device through a stripe-window etched into the substrate contact. Dark spots could be detected in this image for some diodes. These spots rapidly grew larger when the device was loaded with current (dark line defects). They were proven to be associated with dislocations in the crystal; dislocations

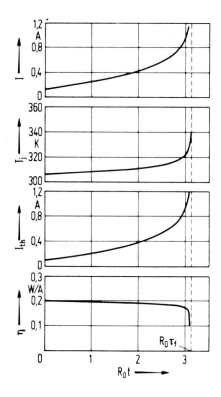

FIGURE 32.   Variation with time of the differential quantum efficiency η_1, threshold current I_{th}, junction temperature T_j, and operating current I needed to maintain a constant optical output of a homogeneously degrading diode. R_o is the initial degradation rate of the threshold current.[40] (After Winstel, G. and Weyrich, C., *Optoelektronik I*, Springer-Verlag, Basel, 1980. With permission.)

become centers of localized heating as a result of nonradiative recombination that causes the dislocations to grow. Dark line defects can be eliminated by a very careful selection of virtually defect-free substrates and a well controlled growth process. In addition to the actual epitaxial growth, every subsequent processing step, such as etching, masking, coating, contacting, etc. is a potential source of defects that can greatly reduce the useful life of the laser.

The principal aging mechanism in a properly processed laser diode is the slow, homogeneous degradation of the mirrors and cavity. Although all the factors affecting this type of degradation are not accurately known, it is sufficient to say that the elimination of the other causes of aging pushed life expectancy figures well over 100,000 hr at room temperature. The variation with time of the principal operating parameters of such a diode can actually be predicted (Figure 32). The most important factor determining useful life expectancy is the initial degradation rate of the threshold current, which means that good life estimates can be obtained after only a few hundred hours of operation, provided some laser parameters such as the temperature coefficient of the threshold current T_o, series resistance, thermal resistance, and the light-current characteristic are known.

Finally, the effect of temperature on aging has to be discussed. Degradation is essentially a thermally induced process. The fraction of the electrical input power that is not converted into light output will eventually reappear as heat in various structural elements of the device. In this, the electroluminescent part of the characteristic below threshold plays the dominant role, since the conversion efficiency of the electroluminescence does not exceed a few percent. Electroluminescent photons that are created in the active region and are not emitted become absorbed in some part of the structure, mainly the substrate, and contribute also to the heating. Thus, the heat loss is mainly determined by the junction voltage and the threshold current. Above threshold, the conversion efficiency rises well above 50%, depending on the design of the optical cavity. In this region, losses are caused by free carrier absorption and nonradiative processes induced by defects. Junction voltage depends on the emission wavelength. Long-wavelength (1.3 to 1.6 μm) devices have a lower junction voltage and are, therefore, in a better position than 0.8 to 0.9 μm devices. Besides conversion losses, one has to deal with Joule heating in the series resistance of the diode (contacts and current flow paths in the device).

The temperature rise inside the diode depends not only on the losses generated in the device, but also on the thermal resistance of the mounting and the design of the heat sink. To obtain a low value of thermal resistance, the diode is often mounted "upside down" on the heat sink, that is the epitaxial layers turned towards the heat sink. This brings the heat producing active layer within a few microns of the heat sink. Since the thickness of the substrate amounts to 100 μm, the "upside up" (that is substrate down) way of mounting introduces a considerably longer heat flow path between sink and active layer. Not all structures allow the "upside down" type of mounting, though many planar diodes do. The disadvantage of this type of mounting is that it requires great care at the assembly because it brings the exit mirrors within a few microns of the heat sink. The way in which the diode is mounted also affects the electrical design of the power supply and drive-circuitry. The substrate is almost always n-type; therefore, "upside down" mounting means that the plus pole of the power supply will be connected to the heat sink, that is to ground. Sometimes the laser is mounted on an electrically insulating silicon wafer placed between heat sink and diode that leaves the choice of grounding to the circuit designer.

Elevated temperatures accelerate the aging process. The temperature dependence of the degradation rate follows an Arrhenius type exponential law ($exp(E_a/kT)$) with an activation energy of about 0.7 eV (Figure 33). Thus operating the diode at 100°C above room temperature accelerates degradation roughly by a factor of a thousand. Elevated temperatures provide, therefore, a useful means to assess the reliability and life expectancy of a diode from a relatively short aging test. Diodes with simple structures, such as the planar oxide stripe laser, can be made to have a very high thermal stability, and a life expectancy of 10^5 hr at 100°C can be achieved.[42] The preconditions for this are low threshold current, high T_o value, and absence of structural defects that can cause rapid degradation.

One can distinguish several current trends in laser diode development. Thermal stability and reliability of long wavelength (1.3 to 1.6 μm) laser diodes are improved continuously. Since optical fibers are practically absorption free in that spectral region, a significant extension of transmission range can be expected from the use of such long wavelength diodes. Movement to shorter wavelengths into the visible region (0.76 to 0.78 μm) is also noticeable, though the prime mover for this development is not the optical communication field but high-density, high-speed data recording. Work is under way to increase power output by pushing mirror-loading limits to higher values. All applications can benefit from this effort. On the technological front, the simultaneous processing of several thousand lasers, including the formation of mirrors by etching could bring significant savings in production costs. Finally, the integration of several optical devices, such as lasers, modulators, drivers, detectors, etc. on a single substrate may bring the same revolution to this field as it did to silicon based integrated electronics.

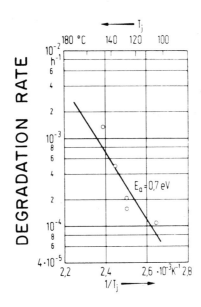

FIGURE 33.   Median degradation rate of laser diodes as a function of junction temperature T_j.[41] (After Winstel, G. and Weyrich, C., *Optoelektronik I*, Springer-Verlag, Basel, 1980. With permission.)

REFERENCES

1. **Bergh, A. A. and Dean, P. T.,** *Light Emitting Diodes,* Clarendon Press, Oxford, 1976.
2. **Kressel, H. and Butler, J. K.,** *Semiconductor lasers and heterojunction LED,* Academic Press, New York, 1977.
3. **Casey, H. C. and Panish, M. B.,** *Heterostructure Lasers,* (Part 5 A and B), Academic Press, New York, 1978.
4. **Pankove, J. I.,** *Optical Processes in Semiconductors,* Prentice Hall, Englewood Cliffs, N.J., 1971.
5. **Winstel, G. and Weyrich, C.,** *Optoelektronik I,* Springer-Verlag, Basel, 1980.
6. **Sell, D. D., Casey, H. C., and Wecht, K. W.,** Concentration dependence of the refractive index for n- and p-type GaAs between 1.2 and 1.8 eV, *J. Appl. Phys.,* 45, 2650, 1974.
7. **Casey, H. C., Sell, D. D., and Panish, M. B.,** Refractive index of $Al_xGa_{1-x}As$ between 1.2 and 1.8 eV, *Appl. Phys. Lett.,* 24, 63, 1974.
8. **Berolo, O. and Wooley, J. C.,** Electroreflectance spectra of $Al_xGa_{1-x}As$ alloys, *Can. J. Phys.,* 49, 1335, 1971.
9. **Casey, H. C., Sell, D. D., and Wecht, K. W.,** Concentration dependence of the absorption coefficient for n- and p-type GaAs between 1.3 and 1.6 eV, *J. Appl. Phys.,* 46, 250, 1975.
10. **Stern, F.,** Calculated spectral dependence of gain in excited GaAs, *J. Appl. Phys.,* 47, 5382, 1976.
11. **Lengyel, G., Nardone, S., and Pommerrenig, D.,** Measurement of electron diffusion length by photoluminescence in p-doped GaAs substrates and epitaxially grown photo cathodes, in *Advances in Electronics and Electron Physics,* Vol. 33 B, Academic Press, London, 1972, 389.
12. **Sze, S. M.,** *Physics of Semiconductor Devices,* John Wiley & Sons, New York, 1969.
13. **Anderson, R. L.,** Experiments on Ge-GaAs heterojunctions, *Sol. St. El.,* 5, 341, 1962.
14. **Ahearn, W. E. and Crow, J. W.,** Linewidth measurements of CW gallium arsenide lasers at 77°K, *IEEE J. Quant. Electron.,* 2, 597, 1966.
15. **Dyott, R. D. and Stern, J. R.,** Group delay in glass-fibre waveguide, *Electron. Lett.,* 7, 82, 1971.
16. **Siemens Corporation,** Semiconductor Components for Optical Communications. Databook of Optoelectronic Semiconductors, Munich, 1980/81.
17. Hitachi CSP Laser Data Sheet, HLN-500, Hitachi America Ltd., Arlington Heights, Ill.
18. **Grosskopf G. and Küller, L.,** Measurement of nonlinear distortions in index- and gain-guiding GaAlAs lasers, *J. Opt. Commun.,* 1, 15, 1980.

19. **Ettenberg, M., Wittke, J. P., and Kressel, H.,** Very high radiance edge-emitting LED, *IEEE J. Quant. Electron.,* 12, 360, 1976.

20. **Paoli, T. L.,** Intrinsic fluctuations in the output intensity of double-heterostructure junction lasers operating continuously at 300 K, *Appl. Phys. Lett.,* 24, 187, 1974.

21. **Jäckel, H.,** Lichtemissionsrauschen und dynamisches Verhalten von GaAlAs-Heterostruktur-Diodenlasern im Frequenzbereich von 10 MHz bis 8 GHz, Dissertation, Nr. 6447, Eidgenössische Technische Hochschule, Zürich, 1980.

22. **Ito, T., Machida, S., Nawata, K., and Ikegami, T.,** Intensity fluctuations in each longitudinal mode of a multimode AlGaAs laser, *IEEE J. Quant. Electron.,* QE-13, 574, 1977.

23. **Pankove, J.,** Temperature dependence of emission efficiency and lasing threshold in laser diodes, *IEEE J. Quant. Electron,* QE-4, 119, 1968.

24. **Carr, W. N.,** Photometric figures of merit for semiconductor luminescent sources operating in spontaneous mode, *Infrared Phys.,* 6, 1, 1966.

25. **Burrus, C. A. and Miller, B. I.,** Small-area, double-heterostructure aluminum gallium arsenide electroluminescent diode sources for optical-fiber transmission lines, *Opt. Commun.,* 4, 307, 1971.

26. **Kressel, H. and Ettenberg, M.,** A new edge-emitting (AlGa)As heterojunction LED for fiber-optic communications, *Proc. IEEE,* 64, 1360, 1975.

27. **Hakki, B. W.,** GaAs double heterostructure lasing behavior along the junction plane, *J. Appl. Phys.,* 46, 292, 1975.

28. **Paoli, T. L.,** Waveguiding in a stripe-geometry junction laser, *IEEE J. Quant. Electron.,* QE-13, 662, 1977.

29. **Streifer, W., Scifres, D. R., and Burnham, R. D.,** Analysis of gain-induced waveguiding in stripe geometry diode lasers, *IEEE J. Quant. Electron.,* QE-14, 418, 1978.

30. **Cook, D. D. and Nash, F. R.,** Gain-induced guiding and astigmatic output beam of GaAs lasers, *J. Appl. Phys.,* 46, 1660, 1975.

31. **Lengyel, G., Wolf, H.-D., and Zschauer, K.-H.,** A study of lateral modes in wide double-heterostructure GaAs-GaAlAs laser diodes, *J. Appl. Phys.,* 49, 1047, 1978.

32. **Kirkby, P. A. and Thompson, G. H. B.,** Channeled substrate buried hetero-structure GaAs-GaAlAs injection lasers, *J. Appl. Phys.,* 47, 4578, 1976.

33. **Tsukada, T.,** GaAs-Ga$_{1-x}$Al$_x$As buried-heterostructure injection lasers, *J. Appl. Phys.,* 45, 4899, 1974.

34. **Tsang, W. T., Logan, R. A., and Ilegems, N.,** High-power-fundamental-transverse mode stripe buried heterostructure lasers with linear light-current characteristics, *Appl. Phys. Lett.,* 32, 311, 1978.

35. **Aiki, K., Nakamura, M., Kuroda, T., Umeda, J., Ito, R., Chinone, N., and Maeda, M.,** Transverse mode stabilized Al$_x$Ga$_{1-x}$As injection lasers with channeled-substrate-planar structure, *IEEE J. Quant. Electron.,* QE-14, 89, 1978.

36. **Kressel, H. and Nelson, H.,** Close-confinement gallium arsenide pn junction lasers with reduced optical loss at room temperature, *RCA Rev.,* 30, 106, 1969.

37. **Hayashi, I., Panish, M. B., Foy, P. W., and Sumski, S.,** Junction lasers which operate continuously at room temperature, *Appl. Phys. Lett.,* 17, 109, 1970.

38. **Alferov, Zh. I., Andreev, V. M., Portnoi, E. L., and Trukan, M. K.,** AlAs-GaAs heterojunction injection laser with a low room-temperature threshold, *Sov. Phys. Semicond.,* 3, 1107, 1970.

39. **Zschauer, K.-H.,** Private communication.

40. **Mettler, K., Wolf, H.-D., and Zschauer, K,-H.,** Calculation of the homogeneous degradation of injection laser parameters from initial degradation rates, *IEEE J. Quant. Electron.,* QE-14, 819, 1978.

41. **Furukawa, Y., Kobayashi, T., Wakita, K., Kawakami, T., Iwane, G., Horikoshi, Y., and Seki, Y.,** Accelerated life test of AlGaAs-GaAs DH lasers, *Jpn. J. Appl. Phys.,* 16, 1495, 1977.

42. **Wolf, H. D., Mettler, K., and Zschauer, K.-H.,** High performance 880 nm (GaAl)As/GaAs oxide stripe lasers with very low degradation rates at temperatures up to 120°C, *Jpn. J. Appl. Phys.,* 20, L693, 1981.

Chapter 5

PHOTODETECTORS

T. P. Lee

TABLE OF CONTENTS

I. INTRODUCTION

This chapter describes the basic principle of operation and performance characteristics of photodetectors used in optical fiber communication systems. The photodetector is an essential element in such systems; it converts the lightwave signals into electrical currents that are subsequently amplified and processed. The ultimate performance requirements are high speed (therefore, large bandwidth) and low-noise (therefore, high sensitivity). The bulk of this chapter is devoted to the discussion of semiconductor photodiodes, both simple PIN diodes and more sophisticated avalanche photodiodes with internal current gain, because they are most compatible with optical fiber systems and have shown best performance thus far.

In the short wavelength region between 0.7 to 0.9 μm wavelength, the discussion is focused on silicon avalanche photodiodes because of their superior performances. Recently, photodetectors for the longer wavelength region from 1.0 to 1.7 μm, where silica-based fibers exhibited low-loss and minimal dispersion, have been developed very rapidly. Among various III—V materials, InGaAs and InGaAsP have emerged as best candidates for the photodiode materials. A considerable portion of the discussion is devoted for a review of this development and to point out present achievements and research interests.

Several previous review articles on photodiodes,[1,2] and on photodetectors in general[3-9] can be useful for supplementary readings.

II. PRINCIPLES OF PHOTODIODES

The basic principle of operation of a photodiode is the conversion of the incident optical signals into electrical currents by the absorption process in the material employed. The performance characteristics of such devices include: (1) photoresponsivity, (2) speed of response, (3) internal current gain, if any, and (4) noise properties. In this section, we shall discuss the fundamental consideration of these characteristics.

A. PIN Photodiodes

1. Photoresponsivity and Quantum Efficiency

In a reverse-biased semiconductor PIN photodiode, as illustrated in Figure 1, incoming photons are absorbed, and electron-hole carrier pairs are generated primarily in the depleted I-region (I stands for intrinsic). The photogenerated carrier pairs are separated by the high electric field in the depletion region and are collected across the reverse-biased junction. As the carriers traverse the depletion region, a displacement current is induced at the load as the signal current. For a semiconductor with absorption coefficient α_o at wavelength λ, the primary photo-current produced by the absorption of incident light of optical power P_o is given by

$$I_p = P_o \frac{q(1 - r)}{h\nu} (1 - e^{-\alpha_o w}) \qquad (1)$$

where q is the electronic charge, $h\nu$ is the photon energy ($h\nu = 1.24/\lambda$ eV, where λ is wavelength in microns), r is the Fresnel reflection coefficient at the semiconductor-air interface, and w is the width of the absorption region. The quantum efficiency η is defined as,

$$\eta = \frac{(\text{number of carriers generated})}{(\text{number of incident photons})}$$

$$= (I_p/q)/(P_o/h\nu) \qquad (2)$$

$$= (1 - r)(1 - e^{-\alpha_o w})$$

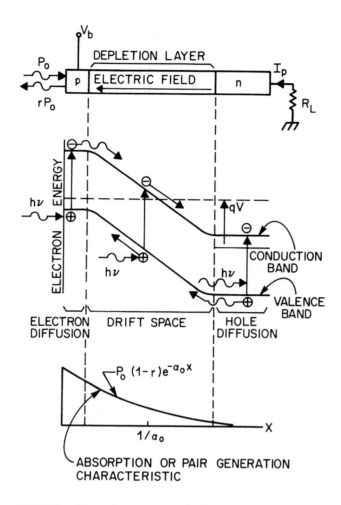

FIGURE 1. Schematic representation showing the principle of operation of a p-i-n photodiode. The energy band diagram under reverse bias, and optical absorption and carrier generation characteristics are shown. (From Melchior, H., *J. Lumin,* 7, 390, 1973. With permission.)

The responsivity R, often used to characterize the photodiode performance, is given by

$$R = I_p/P_o = \eta q/h\nu \qquad (3)$$

For an ideal photodiode ($\eta = 1$), $R = \lambda/1.24$ A/W, where λ is wavelength in microns.

2. Speed of Response

Carriers generated in the depletion region are collected by the junction at a speed equal to the average carrier transit time across the high field depletion region. The velocity of carriers in a semiconductor at a field higher than 10^4 V/cm reaches the scattering limited maximum velocity[10] of about 1×10^7 cm/sec, depending on the carrier types and on the semiconductor materials. Thus the normalized transit time is about 10 psec (10^{-11} sec) per μm of depletion width. Diodes with high quantum efficiency require a depletion width of 2 to 3 μm, for which a maximum response time of 20 to 30 psec can be expected.

Carriers generated outside the high field depletion region, however, have to reach the junction by diffusion, a much slower carrier transport process. The diffusion time is approximately t = w²/2.4 D where D is the diffusion constant of the minority carriers (electron in the p-region, holes in the n-region of the junction). The normalized diffusion time is about 1 nsec (10^{-9}sec)/μm of absorption width in the p-region of the junction in silicon diodes (about 100 psec/μm for III-V semiconductor photodiodes) and several nanoseconds per micron in the n-region of the junction.

The diode capacitance (the junction capacitance and the package capacitance) together with the circuit load resistance gives rise to still another circuit RC time constant relevant to the speed of response of a photodiode. Usually, a 50 Ω resistor is used for the load for high speed operations, and the RC time constant will be 50 psec/pF of total diode capacitance. Neglecting the package capacitance, the junction capacitance is $C_j = \epsilon A/w$, where ϵ is the permittivity of the semiconductor, w is the depletion width, and A is the diode area. For a 250 μm diameter diode, C_j is of the order of 2 to 3 pF, and an RC time constant of a few hundred picoseconds is expected. For fast response, the diode area has to be small.

It is clear, from the above discussion, that the diffusion time is usually the slowest component. In a PIN diode structure the I-region, which is completely depleted, is made to be several penetration depths ($1/\alpha_o$) wide so that insignificant numbers of carriers are generated outside the depletion region and, thus, the diffusion current can be negligible. Conversely, when the depletion width is narrow, as in a pn diode, or when the photons penetrate deep in the bulk region to a distance away from the junction, the diffusion current dominates. As a result, the speed of response is much slower.

3. Absorption Coefficient

The absorption coefficients α_o and penetration depth ($1/\alpha_o$) of commonly used semiconductor materials are given in Figure 2 for the 0.4 to 1.7-μm spectral region. At wavelengths longer than the band-edge wavelength ($\lambda_c = 1.24/E_g$ μm), the absorption decreases sharply. Thus, ideally, a photodiode material should be one with a bandgap energy E_g that is slightly less than the photon energy corresponding to the longest operating wavelength for achieving optimum quantum efficiency and speed, as well as low dark currents (which decreases exponentially with increasing bandgap energy[11]).

In the spectral region from the visible to 1 μm silicon is the preferred material because of its low dark current and its low-noise avalanche gain (see later discussions). Germanium and Indium-Gallium-Arsenide (InGaAs) alloy (and other III—V alloys such as GaAsSb) are suitable for wavelengths to 1.7 μm. One advantage of the alloy material is that its bandgap energy can be varied by adding a fourth element in it and adjusting the relative concentration of the constituents, e.g., $In_{0.7}Ga_{0.3}As_{0.64}P_{0.36}$ with a band edge wavelength near 1.4 μm, as shown in Figure 2.

B. Avalanche Photodiodes (APDs)

When the electric field in the depletion region of a reverse-biased PIN diode is sufficiently high (above 10^5V/cm for Si) the photogenerated primary free carriers (electron-hole pair) can collide with a valence electron with sufficient energy to cause impact ionization of the valence electron into the conduction band, leaving a free hole in the valence band, thus creating an additional carrier pair. The secondary carriers that traverse the high field can in turn be accelerated and cause further ionization, until an avalanche of carriers has been produced.[12] The generation of electron-hole carrier pairs and avalanche multiplication can be described in terms of the ionization rates of electrons and holes that depend upon the electric field strength. The ionization rates, expressed in cm^{-1}, are reciprocal of the average distance a carrier will travel at a given electric field before impact ionization generates a secondary electron-hole pair. The ionization rates for electron and hole, represented by $\alpha(E)$ and $\beta(E)$, in general are not equal, but vary with different semiconductor materials and are sensitive to temperature.[13] The measured ionization rates for a few commonly known semi-

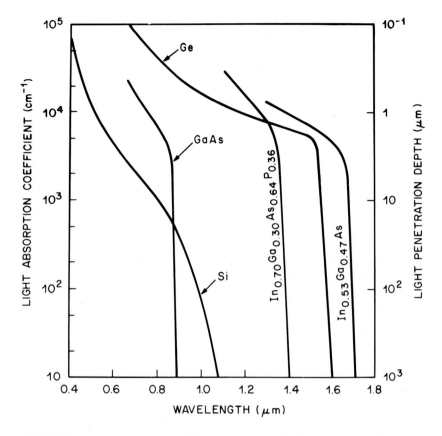

FIGURE 2. Optical absorption coefficient or penetration depth vs. wavelength for silicon, germanium, gallium arsenide, indium gallium arsenide phosphide, and indium gallium arsenide.

conductors are summarized in Figure 3 for Si,[14] Ge,[15,16] $In_{0.14}Ga_{0.86}As/GaAs$,[17,18] InGaAs/InP,[19] and $GaAs_{0.88}Sb_{0.12}$.[20] The ratio of the ionization rates, β/α, and their field dependence have profound effects on the performance of the maximum current gain, the gain-bandwidth product as well as the excess noise due to avalanche process in an APD.

1. Low-Frequency Avalanche (Current) Gain

A conceptual representation of the avalanche process is sketched in Figure 4(a) where the electric field may vary with distance x in the depletion region of width w. The variation of electron current in the depletion region can be written as

$$\frac{d}{dx} J_n(x) = \alpha(x)J_n(x) + \beta(x)J_p(x) + qG(x) \tag{4}$$

where $J_n(x)$ and $J_p(x)$ are electron and hole currents at x, respectively, G(x) is the generation rate of electron-hole pairs by absorption of photons at x and q is the electronic charge. The variation of hole current can be obtained similarly and is given by

$$- \frac{d}{dx} J_p(x) = \alpha(x)J_n(+ \beta(x)J (x) + qG(x) \tag{5}$$

The negative sign appears before $dJ_p(x)/dx$ term because the hole current decreases with increasing x while the electron current increases with increasing x. Thus the total current under DC condition is

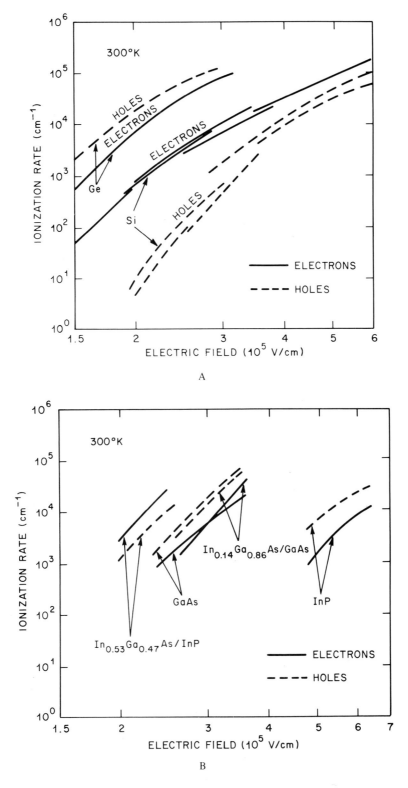

FIGURE 3. Experimentally obtained carrier ionization rates at 300K for (A) silicon and germanium and (B) GaAs, InP, $In_{0.14}Ga_{0.86}As/GaAs$ and $In_{0.53}Ga_{0.47}AS/InP$. (From Lee, T. P. and Li, T., in *Optical Fiber Communications*, Miller, S. E. and Chynometh, Eds., Academic Press, New York, 1979, chap. 18. With permission.)

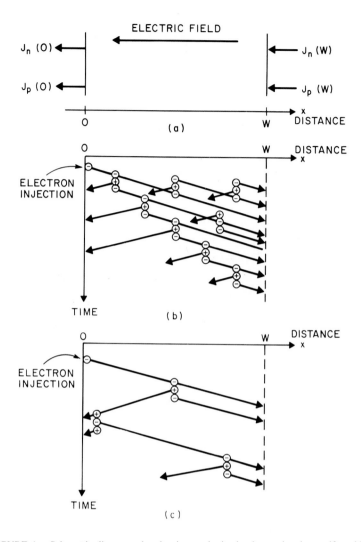

FIGURE 4. Schematic diagrams showing impact ionization by carriers in a uniform high-field region of a semiconductor (a) only electrons undergo ionizing collisions; (b) both carriers undergo ionizing collisions. (From Stillman, G. E. and Wolfe, C. M., *Semiconductors and Semimetals*, Vol. 12, Willardson, P. K. and Beers, A. C., Eds., Academic Press, New York, 1977, 291. With permission.)

$$J = J_n(x) + J_p(x) = \text{const.} \tag{6}$$

If the hole injection at $x = w$ and electron injection at $x = 0$ can be considered separately and the generation in the depletion region can be ignored, the multiplication factors for electron current and hole current can be obtained from Equations 4 and 5 in a closed form,[14] expressed, respectively, by

$$M_n = \frac{J}{J_n(0)} = \frac{\exp\left[\int_o^w (\alpha - \beta)dx\right]}{1 - \int_o^w \beta \exp\left[-\int_x^w (\alpha - \beta)dx'\right]dx} \tag{7}$$

and

$$M_p = \frac{J}{J_p(w)} = \frac{\exp\left[-\int_o^w (\alpha - \beta)dx\right]}{1 - \int_o^w \alpha \exp\left[-\int_o^x (\alpha - \beta)dx'\right]dx} \tag{8}$$

The solutions to Equations 7 and 8 are complicated by the fact that both α and β are functions of position. However, solutions of M_n and M_p for two special cases, i.e., (1) β (or α) = 0, and (2) $\beta = \alpha$, both independent of position (uniform field), can be much simplified. The former case, $\beta = 0$, corresponds to the most ideal avalanche process while the latter case, $\alpha = \beta$, corresponds to the most non-ideal avalanche process.

For $\beta = 0$ and uniform field, Equation 7 can be reduced to

$$M_n = \exp\left[\int_o^w \alpha dx\right] = \exp(\alpha w) \tag{9}$$

Equation 9 implies that M_n increases exponentially with the number of ionizing carriers (αw) in the high field region with width w and that there is no sharp beakdown. Figure 5(b) depicts the build-up of the avalanche process with time under $\beta = 0$ condition. The total current pulse starts with the initial electron injection at x = 0, increases during the time of electron transit through the high field region, then reduces to zero in about the hole-transit time. The pulse width is approximately the average of electron and hole transit times and is independent of the amount of multiplication. Thus there is no gain-bandwidth product limit, and the bandwidth is determined only by the carrier transit time. It can also be said, from Equation 9, that for large M_n there exists a large number of ionizing carriers (αw) within the depletion region w at a given time. Thus a statistical variation of the impact ionization process will cause only a small fluctuation in the total number of carriers, and the avalanche process will produce small excess noise.

However, the situation changes drastically when both carriers ionize, i.e., $\beta = \alpha$. Again assuming uniform field, Equations 7 and 8 become

$$M_n = M_p = \frac{1}{1 - \int_o^w \alpha dx} = \frac{1}{1 - \alpha w} \tag{10}$$

A sharp breakdown, defined as $M_n \to \infty$, occurs when the voltage corresponding to the condition $\alpha w = 1$ is reached. This implies that avalanche breakdown occurs when, on the average, each injected carrier produces one electron-hole pair during its transit through the high-field depletion region. Figure 4(c) shows a schematic diagram of the avalanche gain build-up with time, where the the injected electron at x = 0 initiates the ionization. The secondary holes meanwhile traverse in the $-x$ direction and generate further electron hole-pairs, producing a positive feedback mechanism. The larger the gain, the longer this process persists. Thus there is a gain-bandwidth product relationship. It is also clear that for a large gain, there are few ionizing carriers in the high field region. Thus statistical variations in the impact ionization process will produce large fluctuations in the gain and cause considerable excess noise.

In a practical photodiode, both the electrons and holes contribute to impact ionization, and in general, $\alpha \neq \beta$. If a uniform field is assumed again, the multiplication of electrons injected into the high field region at x = 0 can be written as

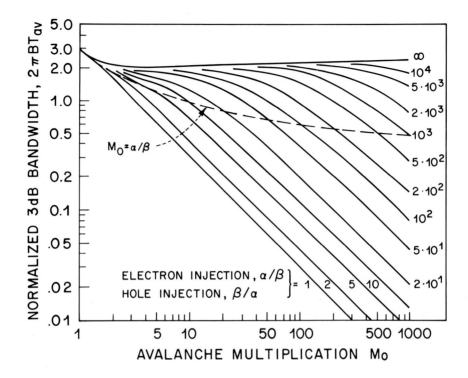

FIGURE 5. Theoretical 3-dB bandwidth B times $2 \pi \tau_{av}$ (τ_{av} = average carrier transit time) of an avalanche photodiode plotted as a function of the low-frequency multiplication factor M_o, for various value of α/β for electron injection (or β/α for hole injection). Above the dashed curve ($M_o = \alpha/\beta$) bandwidth is nearly independent of gain; below the dashed curve a constant gain bandwidth product applies.[1,23]

$$M_n = \frac{[1 - (\beta/\alpha)] \exp\{\alpha w[1 - (\beta/\alpha)]\}}{1 - (\beta/\alpha) \exp\{\alpha w[1 - (\beta/\alpha)]\}} \tag{11}$$

Equation 11 can be simplified further, for small values of β/α, as

$$M_n \cong M_l/[1 - (\beta/\alpha)M_l] \tag{12}$$

where $M_l = \exp[\alpha w]$ as given by Equation 9. The expression of Equation 12 is similar to the gain of a positive feedback amplifier with feedback factor (β/α) and an open-loop gain M_l.

The low-frequency gain is conveniently written as a function of bias voltage,[15] as

$$M_o = \frac{I}{I_p} = \frac{1}{1 - (V_j/V_B)^n} \tag{13}$$

where I is the multiplied diode current, I_p is the primary current, V_B is the breakdown voltage, V_j is the effective junction voltage, and n is a fitting factor depending on the diode material and structure. At large gain, the voltage drop due to the diode series resistance and load resistance must be taken into account, and Equation 13 becomes[21]

$$M_o = \frac{I}{I_p} = \frac{1}{1 - \left[\dfrac{V - IR}{V_B}\right]^n} \tag{14}$$

where V is the bias voltage and IR is the voltage drop. For $V_B >> IR$, Equation 14 can be approximated by[21]

$$M_o = \frac{I}{I_p} = V_B/nIR \tag{15}$$

from which the maximum value of the multiplication factor is derived

$$M_o(max) = (V_B/nI_pR)^{1/2} \tag{16}$$

Since I_p includes both the primary photogenerated current and the primary thermally excited dark current, it can be seen that the primary dark current will set a limit on the value of the maximum gain achievable. For example, maximum gains as high as 10,000 have been observed in Si APDs,[22] but much smaller gains ($M \sim 200$) occur in Ge APDs[21] due to the large dark current.

2. Gain-Bandwidth Product

Figure 5 shows the calculated bandwidth for an ideal avalanche photodiode having an avalanche region of uniform electric field in which the drift velocities of electrons and holes are assumed to be equal.[23] The 3-dB bandwidth B, normalized to $2\pi\tau_{av}$ where τ_{av} is the transit time, is plotted as a function of the low-frequency gain M_o with α/β as a parameter. The dashed curve is for $M_o = \alpha/\beta$. Above this curve where $M_o < \alpha/\beta$, the bandwidth is largely independent of gain. Below the curve where $M_o > \alpha/\beta$, the curves are almost straight lines, indicating a constant gain-bandwidth product. The high-frequency multiplication factor can be approximated by

$$M(\omega) = \frac{M_o}{[1 + (\omega M_o \tau_{eff})^2]^{1/2}} \tag{17}$$

where τ_{eff} is approximately equal to $N(\beta/\alpha)\tau_{av}$, N varies from 1/3 (at $\beta/\alpha = 1$) to 2 (at $\beta/\alpha = 10^{-3}$), and ω is 2π times the frequency. Equation 17 implies a constant gain-bandwidth product:

$$M_o B = (\alpha/\beta)/N\tau_{av} , M_o > \alpha/\beta \tag{18}$$

In practice, τ_{eff} is also structure dependent and this effect can be taken into account by letting $\beta/\alpha = k_{eff}$ defined in the next section.[24,25]

3. Avalanche Multiplication Noise

The multiplication factor M_o given in the above section represents the average gain of the injected electron or hole current. However, because of the statistical nature of the multiplication process, not all photogenerated carriers experience the same gain. There is a gain distribution that gives rise to the noise due to the multiplication process.[26-30] For a diode with uniform electric field, as in a p—i—n diode, the shot noise in the multiplied diode current is given by

$$\langle i_n^2 \rangle = 2q(I_p + I_d)\langle M^2 \rangle B \tag{19}$$

where $\langle i_n^2 \rangle$ is the mean square noise current, I_p is the primary photocurrent, I_d is the primary dark current, q is the electronic change, B is the effective noise bandwidth and $\langle M^2 \rangle$ is the mean square value of the multiplication factor (gain). Because the gain fluctuates, the mean square value $\langle M^2 \rangle$ is greater than the square of the mean $\langle M \rangle^2 = M_o^2$; the excess noise can be characterized by the excess noise factor $F(M_o) = \langle M^2 \rangle / M_o^2$. Therefore Equation 19 can be written as

$$\langle i_n^2 \rangle = 2q(I_p + I_d)M_o^2 F(M_o)B \tag{20}$$

The excess noise factor is dependent on the ratio of the ionization rates, α/β, and on the DC multiplication factor M_o. When both carrier species produce impact ionization (Figure 4c), a small statistical variation in the regenerative avalanche process can cause a much larger fluctuation in gain, as compared to the case in which only one type of carrier ionizes (Figure 4b). Thus either $\alpha = 0$ or $\beta = 0$ minimizes noise.

Expressions for the excess noise factor have been derived for both electron injection and hole injection.[26,27] For electron injection alone, the simplified expression for F can be written as

$$F = kM_o + \left[2 - \frac{1}{M_o}\right](1 - k) \tag{21}$$

where $k = \beta/\alpha$ is assumed to be constant throughout the avalanche region. For hole injection alone, the above expression for F still applies if k is replaced by $k' = \alpha/\beta$.

Of particular interest are two special cases: (1) only electrons cause ionizing collisions, i.e., $\beta = 0$, and (2) both carrier species ionize, with $\alpha = \beta$. In case (1) $F = 2$ for large M_o, and in case (2) $F = M_o$.

In practical avalanche photodiodes the electric field in the avalanche region of width w is not uniform; therefore, the impact ionization rates of the carriers must be weighted accordingly to give[27]

$$k_1 = \int_o^w \beta(x)M(x)dx \bigg/ \int_o^w \alpha(x)M(x)dx \tag{22}$$

and

$$k_2 = \int_o^w \beta(x)M^2(x)dx \bigg/ \int_o^w \alpha(x)M^2(x)dx \tag{23}$$

The excess noise factors for electron injection and hole injection become,[27,31]

$$F_e = k_{eff}M_e + \left[2 - \frac{1}{M_e}\right](1 - k_{eff}) \tag{24}$$

and

$$F_h = k'_{eff}M_h - \left[2 - \frac{1}{M_h}\right][k'_{eff} - 1] \tag{25}$$

where $k_{eff} \simeq k_2$, $k'_{eff} = k_2/k_1^2$, and the subscripts e and h denote electron and hole, respectively.

Figure 6 shows F_e as a function of M_e with k_{eff} as a parameter. It is seen again that a small value of k_{eff} is desirable to minimize excess noise.

When light is absorbed on both sides of the junction so that both electrons and holes are

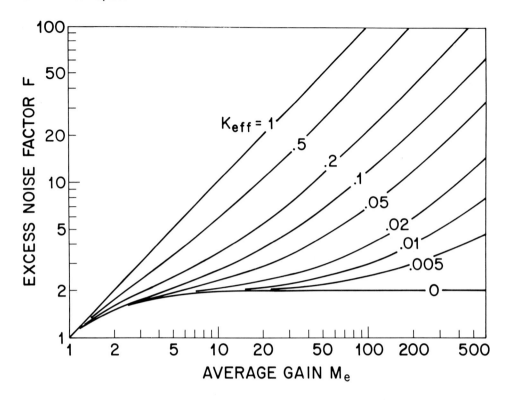

FIGURE 6.   Excess noise factor as a function of the low-frequency gain for electrons, $\dot{M}_e$, with κ_{eff}, the ratio of the weighted ionization rates, as a parameter.[1,31]

injected into the avalanche region additional noise is introduced;[31,32] the effective noise factor is given by

$$F_{eff} = \frac{fM_e^2F_e + (1 - f)M_h^2F_h}{[fM_e + (1 - f)M_h]^2} \qquad (26)$$

where the injection ratio $f = I_{no}/(I_{po} + I_{no})$, and I_{no} and I_{po} are the injected electron and hole currents, respectively. The effective excess noise factor F_{eff} for a diode having a k_{eff} of 0.005 is given in Figure 7 as a function of the average gain, $\overline{M}_o = fM_e + (1 - f)M_h$, with the injection ratio f as a parameter. It clearly shows that the excess noise factor is both material- and structure-dependent. Hence in order to attain low noise, not only k_{eff} must be small, but also the incident light must be absorbed on the correct side of the junction.

The above discussions on avalanche photodiodes may be summed up as follows: to achieve low-noise and wide-bandwidth in an avalanche photodiode, it is necessary that the impact ionization rates of the carriers be as different as possible and that the avalanche process be initiated by the carrier species with the higher ionization rate.

III. PRACTICAL PHOTODIODES

A. Photodiodes for Short Wavelength—Silicon Avalanche Photodiodes

In the wavelength region between 0.4 to 1.0 μm, silicon photodiodes are preferred because of the low dark current and a large ratio of carrier ionization rates. Thus, silicon avalanche photodiodes with high sensitivity, wide bandwidth, and low noise have been produced.

In order to take full advantage of the possibility offered by silicon for carrier multiplica- with very little excess noise, a "reach-through" structure was proposed and imple-

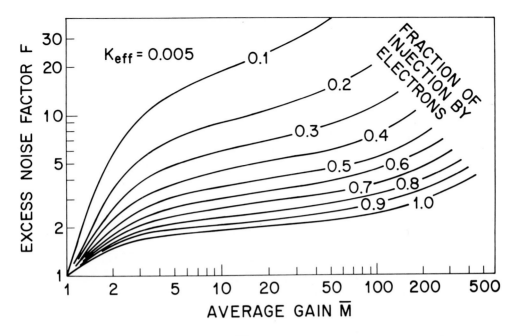

FIGURE 7.   Excess noise factor vs. average gain $\overline{M}$ for mixed injection with $\kappa_{eff} = 005$. The plot illustrates the importance of initiating the avalanche process by the carrier species with the higher ionization rate.[1,31]

mented.[31-39] The ''reach-through'' structure is composed of p^+—π—p—n^+ layers as illustrated in Figure 8(A). The high-field p—n^+ junction, where electron-initiated avalanche multiplication takes place, is formed by diffusion or ion-implantation with precise doping concentration. Under low reverse bias, most of the voltage drop appeared across the p—n^+ junction. As the bias is increased, the depletion layer widens predominantly into the p region and, at a certain voltage V_{rt} below the breakdown voltage of the p—n^+ junction, *reaches through* to the nearly-intrinsic π region. The applied voltage in excess of V_{rt} is dropped across the entire π region. Since the π region is much wider than the p region, the field in the multiplication region and, therefore, the multiplication factor will increase relatively slowly with increasing voltage above V_{rt}. In the operating range, the field in the π region is substantially lower than that in the p—n^+ junction, but is high enough to maintain limiting carrier velocities, thus assuring a fast speed of response.

Because the light incident on the p^+ surface is almost completely absorbed in the π region, a relatively pure electron current is injected into the high-field p—n^+ junction where carrier multiplication takes place. Thus a nearly ideal situation is obtained where the carrier type with the higher ionization rate initiates the multiplication process, resulting in current gain with very little noise. An excess noise factor of 4 at a gain of 100 (corresponding to $k_{eff} = 0.02$) has been observed in a silicon reach-through avalanche photodiode (RAPD) produced for operation in the 0.8 to 0.9 μm wavelength range.[34] In order to take advantage of the larger difference in carrier ionization rates at lower fields in silicon a device has been made in which the p—n^+ avalanche region is formed by a combination of ion implantation and epitaxy; extremely low excess noise corresponding to $k_{eff} \simeq 0.008$ to 0.014 has been obtained.[40]

The epitaxial structure shown in Figure 8(B) is more amenable to fabrication on large silicon wafers with good control of the doping profile. The wafer construction consisting of n^+—p—π—p^+ layers is formed by ion implantation and diffusion on π-type epitaxial material grown on a p^+ silicon substrate.[36] The incident light now enters the structure through the n^+ contacting layer on the surface of the epitaxial material. The increase in excess noise

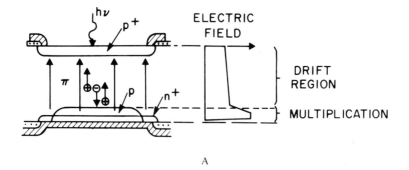

SILICON n^+-p-π-p^+
AVALANCHE PHOTODIODE

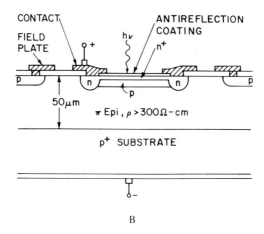

FIGURE 8.   Construction of avalanche photodiodes: (A) Silicon p^+—π—p—n^+ reach-through structure — the electric fields in the drift and multiplication regions are shown on the side;[1] (B) An epitaxial silicon reach-through avalanche photodiode with n^+—p—π—p^+ structure. Diameter of the light-sensitive high-gain region is 100 μm. (From Melchior, H., Hartman, A. R., Schinke, D. P., and Seidel, T. E., *Bell Syst. Tech. J.*, 57, 1791, 1978. With permission.)

caused by the resulting generation and mixed-injection of both carrier types into the avalanche region can be minimized by tailoring the field profile so that electrons injected into the region encounter a lower field than the holes generated within and in front of the region; thus fluctuations due to combined electron- and hole-initiated multiplication processes are minimized. The observed excess noise factor for such a device is 5 at an average gain of 100, which compares very favorably with the value of 4 obtained under the nearly ideal light-injection conditions discussed earlier.[34,36]

A curve of the measured quantum-efficiency (without avalanche gain) of these devices from visible wavelengths to 1.06 μm is shown in Figure 9; near 0.8 μm the efficiency is close to 100%. The current gain as a function of reverse bias voltage at different temperatures is given in Figure 10 for illumination at λ = 0.825 μm. In the range of bias voltage shown, the onset of gain occurs at −60 V, and avalanche breakdown occurs from about −250 to −400 V. The response has been measured using a 0.22-nsec laser pulse at λ = 0.838 μm. The observed duration (FWHM)* of the multiplied output current pulses is about 10 nsec

*   Full Width at Half Maximum.

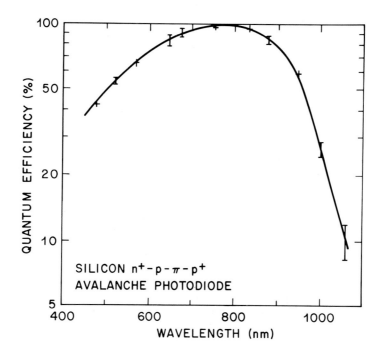

FIGURE 9. Measured quantum efficiency of the silicon n$^+$—p—π—p$^+$ RAPD vs. wavelength. (From Melchior, H., Hartman, A. R., Schinke, D. P., and Seidel, T. E., *Bell Syst. Tech. J.*, 57, 1791, 1978. With permission.)

at the onset of gain, but decreases very rapidly as the π region is depleted (complete depletion occurs at -100 V), to slightly less than 1 nsec at high gain (M $=$ 100).

Avalanche carrier multiplication has been observed to be uniform over the center part of the light-sensitive area of the diodes. (At a gain of 100 the gain is uniform within $\pm 10\%$ over a diameter of 80 to 100 μm.) The dark currents at room temperature are in the low 10^{-11} A range and depend only slightly on bias voltage. The component that is generated in the bulk and is multiplied is estimated to be in the low 10^{-13} A range. The measured variation of the excess noise factor with current gain at $\lambda = 0.8$ μm is presented in Figure 11. The excess noise also varies with wavelength and is less at longer wavelengths (due to increased injection ratio for electrons).

As shown in Figure 10, the current gain of the avalanche diode varies considerably with temperature, especially at high gain.[41] For practical operation, stabilization of gain or compensation for gain variations is necessary. Many schemes are possible. A method that incorporates the required compensation in the automatic-gain-control feedback loop has been implemented in a 45-Mb/sec optical repeater.[42]

In an effort to increase the speed of response of a n$^+$—p—π—p$^+$ reach-through avalanche photodiode without increasing the operating voltage, a build-in field in the light-absorption region was provided by grading the doping profile so as to increase the carrier velocities.[38] Another scheme involved the use of a thin-film structure with a highly reflecting back contact to increase the internal quantum efficiency of a relatively narrow depletion region.[43] Response times of 150 to 200 psec have been achieved in both cases.

For certain optical-fiber applications, an array of closely packed photodiodes may be desirable. A monolithic linear array of ten silicon avalanche photodiodes fabricated on a single chip displayed gain variations of $\pm 2\%$ from diode to diode and an isolation of -57 dB between adjacent diodes.[44] Another array consisting of PIN photodiodes coupled to optical waveguides integrated on a single silicon chip was also demonstrated.[45] Using eva-

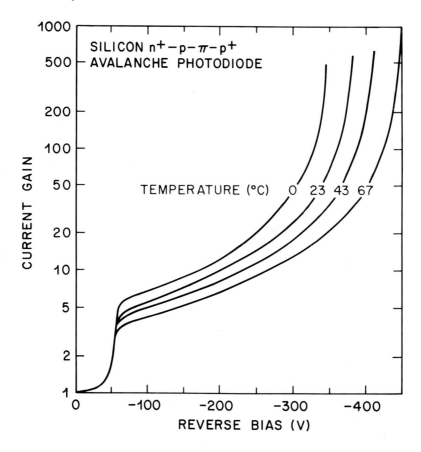

FIGURE 10.   Current-gain vs. reverse bias characteristic of the silicon n^+—p—π—p^+ RAPD at different ambient temperatures measured with illumination at $\lambda = 0.825$ μm. (From Melchior, H., Hartman, A. R., Schinke, D. P., and Seidel, T. E., *Bell Syst. Tech. J.*, 57, 1791, 1978. With permission.)

nescent-wave coupling between waveguides and diodes, device quantum efficiencies of ~80% were obtained at $\lambda = 0.633$ μm.

B. Photodiodes for Long Wavelength

In the long wavelength region between 1.0 to 1.7 μm where silica-based optical fibers exhibited lower loss, germanium and a few III—V semiconductor alloys are suitable materials for photodiodes, as can be seen from the optical absorption characteristics given in Figure 2.

The absorption edge of germanium is near 1.6 μm at room temperature, so its absorption coefficient is large ($\gtrsim 10^4 \text{cm}^{-1}$) over the entire wavelength range of interest for optical-fiber applications. Naturally germanium is a candidate material for detectors at wavelengths beyond 1 μm where the response of silicon falls off. However, the measured value of the ratio of the carrier ionization rates β/α is only about two (see Figure 3), implying a relatively high excess noise factor for avalanche multiplication (F $\simeq$ M/2). In addition, because of the narrower bandgap, the bulk (diffusion-dependent) dark current that undergoes multiplication is expected to be much higher than in silicon. The dark-current problem is further aggravated by the lack of a well-developed surface-passivation technology for the material; thus surface leakage currents tend to be very high and unstable, and play an important role. These shortcomings notwithstanding, devices with usable sensitivity and fast response have been made.

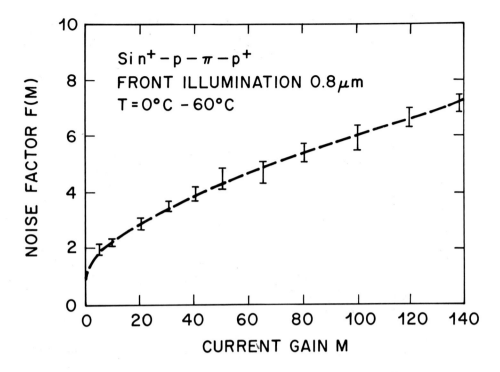

FIGURE 11. Excess noise factor F(M) of the silicon N^+—p—π—p^+ RAPD vs. low-frequency current gain M for illumination at $\lambda = 0.8$ μm. (From Melchior, H., Hartman, A. R., Schincke, D. P., and Seidel, T. E., *Bell Syst. Tech. J.*, 57, 1791, 1978. With permission.)

Various III—V semiconductor alloys, which are being studied for use as optical-source materials, are also under active investigation for use as detector materials for longer-wavelength applications. An attractive feature of these alloys is that their bandgaps depend upon composition; hence it is possible to optimize detector performance by choosing a composition that places the absorption edge just above the wavelength of operation, thereby ensuring high quantum efficiency and speed of response with low dark current. The measured ratio of ionization rates ($\beta/\alpha = 2$ to 3) in these alloy materials, however, is not appreciably different from that of germanium. Recent studies have shown that the impact ionization rate of holes can be appreciably larger than that of electrons because of the resonant split off valence bands,[46,47] and a value of $\beta/\alpha = 20$ has been observed in $Ga_{1-x}As_xSb$ near 1.4 μm wavelength.[48] These findings have implications for other related alloy materials and may lead to interesting possibilities in the search for a suitable longer-wavelength photodetector material.

Experimental photodiodes have been made using various III—V alloys, including GaAs, AlGaAs, InGaAs, GaAsSb, GaAlSb, and InGaAsP. Among these, photodiodes made in InGaAs and InGaAsP have emerged as the most successful ones, and have been employed in many experimental long wavelength fiber systems. Considerable research interest at present is focused on ternary- and quaternary-alloy materials and photodiodes.

1. Germanium Photodiodes

Both high speed germanium PINs[49] and APDs[21,50] have been made for some time, and recently improvements have been made in reducing the dark current and decreasing the excess noise.[51-53] Typical dark currents that undergo multiplication are 50 nA to 100 nA at room temperature and up to ten times higher at the maximum operating temperatures of many fiber systems. In spite of these achievements, the dark currents will limit the amount of useful avalanche gain, which in turn reduces much of the advantage of APDs. The use of germanium APDs would appear limited to extremely high bit-rate systems where large dark currents are permissible[54] or to applications where the devices can be cooled.

2. InGaAs and InGaAsP Photodiodes

a. PIN Photodiodes

The ternary $In_{0.53}Ga_{0.47}As$ and the quaternary $In_xGa_{1-x}As_yP_{1-y}$ semiconductor alloys that have been used extensively as materials for long wavelength light sources are also most useful detector materials.[55-61] Lattice matched to an InP substrate, the ternary $In_{0.53}Ga_{0.47}As$ has a long-wavelength cutoff at 1.65-μm (Eq = 0.75 eV) at room temperature. Thus, as in Ge, this ternary is capable of detecting photons from the visible to 1.7 μm wavelength. For this reason most recent detectors are made of ternary material. However, because of the larger bandgap, diodes made of the quaternary tend to have lower dark current.[60] The construction of a back-illuminated mesa structure InGaAs PIN photodiode[61] is shown in Figure 12(A). An n-type InGaAs layer was epitaxially grown on an InP substrate, also n-type, by liquid phase epitaxy. A p—n junction can be formed either by growth or by diffusion of Zn. Mesas were formed by chemical etch to a diameter of 75 to 100 μm, large enough to be compatible with most fiber dimensions, yet small enough to have low diode capacitance and to contain a minimum number of material defects in the junction that may cause premature breakdown (microplasma breakdown).

The photoresponsivity of this device is shown in Figure 12(B). The responsivity ($\sim$0.6A/W) is almost flat over the entire wavelength region of interest as a result of the back-illumination structure that takes advantage of the transparency of the InP substrate at long wavelengths, and utilizes the fact that the low recombination rate of the photogenerated carriers at the InGaAs-InP interface improves the quantum efficiency at shorter wavelengths. The highest external quantum efficiency at 1.3 μm wavelength was 70%, with typical values of 55 to 60% without antireflection coatings. The lowest leakage current at room temperature was below 1 nA, with typical values below 10 nA, at $-$10 V bias. It increased by about a factor of ten at 70°C. At higher voltages, however, the dark current increased exponentially with voltage. It has been shown that this characteristic is due to tunneling[62-64] that is more pronounced in materials, such as InGaAs, with small bandgaps and small effective electron masses. For minimizing tunneling current at voltages below 50 V, the impurity concentration in the n-type ternary layer has to be maintained below 2×10^{15} cm^{-3}. The total diode capacitance (including package capacitance) was 0.3 to 0.5 pF, resulting in a speed of response better than 100 psec. When the diode capacitance was further reduced to less than 0.05 pF by use of a small mesa (18 to 25 μm diameter) mounted directly onto a stripeline circuit, as shown in Figure 13, an ultrafast response time, limited only by the transit time of 30 psec, has been reported.[65]

Similar mesa-structure InGaAs PIN photodiodes were made in ternary materials grown by molecular beam epitaxy.[66] Planar structures were also made using SiN as the diffusion mask as well as the passivation layer.[67-69] These devices, however, exhibited larger dark currents due to surface leakage. For mesa structures surface passivation over the junction edge by use of silicon resin[61] or polyimide film[70] was found to be necessary; the subject of surface passivation is not well understood and is under further investigation.

By growing layers of different compositions on the same InP substrate, and by forming p—n junctions in the respective layers, dual-wavelength detectors, capable of detecting separately (and therefore demultiplexing) signal photons from two wavelength bands in the 1.0 to 1.6 μm region (as in a wavelength multiplexed system) have been demonstrated.[71-74] The construction of this device is illustrated in Figure 14A. The wafer consisted of a Q_1 layer ($In_{0.70}Ga_{0.30}As_{0.66}P_{0.34}$, E_g = 1.0 eV) that detects photons in the 0.95 to 1.25 μm wavelength region, and a Q_2 layer ($In_{0.53}Ga_{0.47}As$, E_g = 0.75 eV) that detects photons

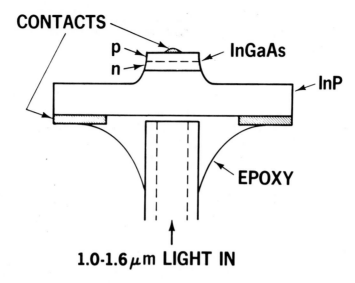

A

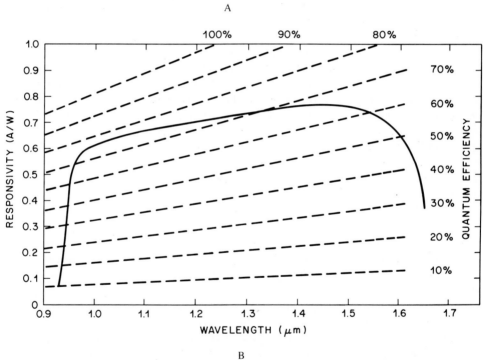

B

FIGURE 12. The blackilluminated $In_{0.53}Ga_{0.47}As/InP$ p-i-n photodiode for the 1.0 to 1.6 μm wavelength region: (A) a cross-sectional view of the diode construction; (B) the photo-response of the diode without anti-reflection coating. (From Lee, T. P., Burrus, C. A., and Dentai, A. G., *IEEE J. Quantum Electron.*, QE-15, 30, 1979. With permission.)

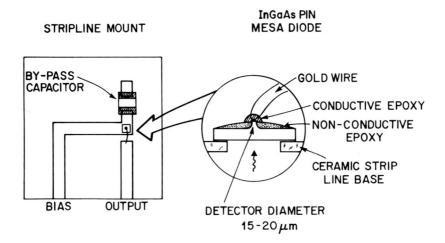

STRIPLINE MOUNT

InGaAs PIN
MESA DIODE

BY-PASS CAPACITOR

GOLD WIRE

CONDUCTIVE EPOXY

NON-CONDUCTIVE EPOXY

CERAMIC STRIP LINE BASE

BIAS   OUTPUT

DETECTOR DIAMETER
15-20 μm

FIGURE 13.   An ultra-fast $In_{0.53}Ga_{0.47}As/InP$ p-i-n photodiode in strip-line mount. The insert shows the construction and mounting of the diode chip with a 18 to 25 μm diameter mesa.

in the 1.25 to 1.65 μm region. An InP barrier layer between Q_1 and Q_2 layers was necessary to block diffusion of the photogenerated carriers between the adjacent junctions, thereby reducing the cross talk between the two channels. Figure 14B shows the photo-response of the detector. The quantum efficiency was 50% at 1.15 μm and 65% at 1.3 μm. The cross talk levels were − 30 dB at 1.15 μm and − 43 dB at 1.3 μm. Dual-wavelength receivers employing such devices were reported also.[72,74]

c. Avalanche Photodiodes

As mentioned above, in a homojunction device, the dark current increases exponentially with reverse bias when the voltage approaches avalanche breakdown. The excessive dark current has been found to be tunneling current under reverse bias that is more profound in materials with small bandgaps. To reduce the tunneling current, a structure that separates the light absorbing region from the multiplying region was proposed.[75-83] In such a structure, the absorption occurs in the narrow bandgap material, whereas the multiplication takes place in the high field region in the wide bandgap material. Both mesa and planar constructions, as shown in Figure 15, have been reported. With proper doping of the n—InP and n—InGaAs (or InGaAsP) layers,[79] the electric fields can be sufficiently high to produce avalanche gain in the InP layer and still be low enough in the ternary layer to avoid tunneling. Average gains between 10 and 60 have been obtained with reasonable unmultiplied dark currents—tens of nanoamperes for InGaAsP devices and few hundred nanoamperes for InGaAs devices. Maximum gain of several hundred can be obtained at much higher dark currents. For low-noise receiver applications, however, the best excess noise factor F was 3 at M = 10 for InGaAsP APDs, but increased to 6 at M = 20, implying a varying K_{eff} value from 0.2 to 0.4. An empirical relationship[81] F = 0.42M was found for 5 ≤ M ≤ 35. At M > 50, F ≅ M. The excess noise factor for InGaAs APDs was similar to that described above.[78,84] The measured excess noise factor F vs. the multiplication M is shown in Figure 16.

3. AlGaAsSb Photodiodes

Photodiodes made in AlGaAsSb, grown lattice-matched to GaSb substrate, with bandgap energies for the 1.3 to 1.6 μm spectral range have been reported.[48,85-88] Although relatively high quantum efficiency and moderate gain (M ≅ 50) have been obtained, the leakage currents, believed to be surface related, are excessive. An interesting feature of this material

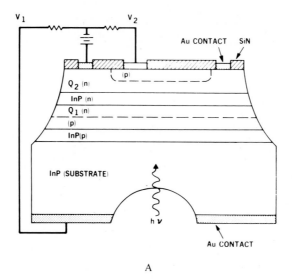

A

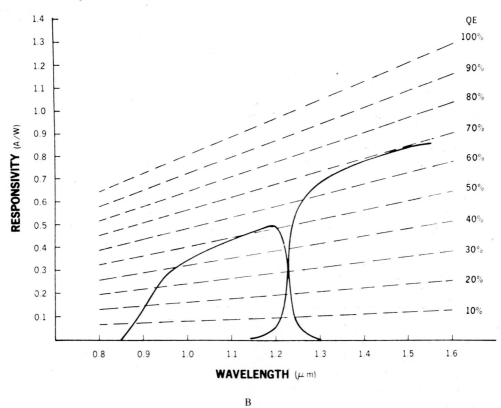

B

FIGURE 14. The dual-wavelength demultiplexing photo detector: (A) construction of the device showing two active quaternary layers separated by an InP layer; (B) the superimposed photo-response of each diode. (From Campbell, J. C., Dentai, A. G., Lee, T. P., and Burrus, C. A., *IEEE J. Quantum Electron.*, QE-16, 601, 1980. With permission.)

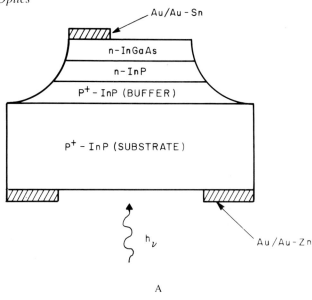

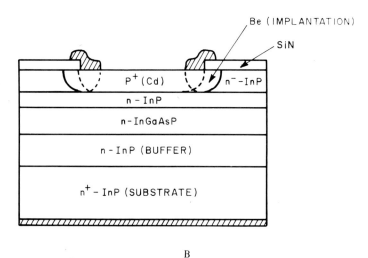

FIGURE 15.   InGaAs/InP avalanche photodiodes: (A) a mesa structure APD —
long-wavelength photons are absorbed by the InGaAs layer, where the photogenerated carriers are multiplied in the high-field region in the InP p⁺-n junction;
(B) a planar construction of the InGaAsP APD. (From Shirai, T., Osaka, F.,
Yamasaki, S., Nakajima, K., and Kaneda, T., *Electron Lett.*, 7, 826, 1981. With
permission.)

system is that the hole ionization rate exhibits a resonant characteristic as a function of the
alloy composition.[48] This resonance, attributed to impact ionization initiated by holes from
the split-off valence band, occurs strongly in material with a bandgap corresponding to 1.4
μm wavelength. As a result, it produced a large hole to electron ionization rate ratio β/α
≈ 20. This high ratio is potentially useful for detectors with low excess noise, provided the
dark current can be kept small.

4. *Other Photodetectors and Future Developments*

In addition to PIN and APD photodetectors, other detectors such as phototransistors,[89-93]
photoconductors,[94-96] pnpn detectors,[97] and integrated PIN-FET[98] have been reported.

A new avalanche photodiode structure that uses a multi-quantum-well of alternating wide

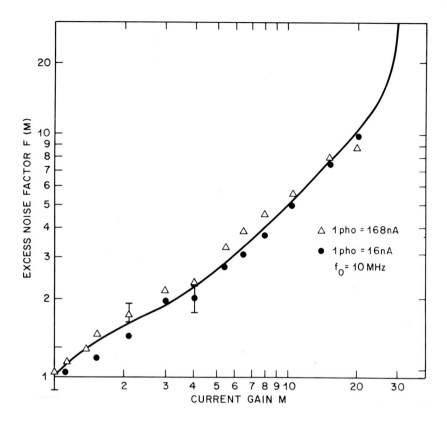

FIGURE 16. Measured excess noise factor vs. low-frequency current gain of a mesa-type InGaAs/InP avalanche photodiode. (From Forrest, S. R., Williams, G. F., Kim, O. K., and Smith, R. G., *Electron Lett.*, 7, 917, 1981. With permission.)

and narrow gap materials has been proposed.[99] In this structure the electrons, upon leaving the wide-bandgap layer and entering the narrow-bandgap layer, have an energy in excess of ionization energy in the narrow gap material. This effect gives rise to an enhanced electron ionization rate. Experimental devices, using alternating layers of $Al_{0.45}Ga_{0.55}As$ (550Å) and GaAs (450Å) grown to a total thickness of 2.5 μm by molecular beam epitaxy were fabricated.[100] An improvement of $\alpha/\beta = 10$ was demonstrated. These super lattice materials and structures, as well as the aforementioned resonance effect of the split-off valence bands to enhance the ionization rate of holes for the reduction of the excess noise of avalanche photodiodes, are areas of current research.

IV. RECEIVER PERFORMANCES

The performance of a receiver in an optical system depends not only on the properties of the photodiode but also on the characteristics of the amplifier that follows the photodiode. Detailed consideration of optical receiver design can be found elsewhere.[54,101,102] We present here a summary of the results achieved with photodiodes described in previous sections.

In the 0.8 to 0.9 μm spectral region, Si-APDs are employed in most receivers to achieve the ultimate sensitivity. Bipolar amplifiers often have been used to obtain large dynamic range ($\cong$40 dB) requiring only moderate equilization. In Figure 17, we present a summary of the receiver performance in the form of a chart. The ordinate is the received average optical power[*] at 10^{-9} bite-error-rate (BER), and the abscissa is the bit-rate of the digital

---

[*] Assume 50% duty cycle.

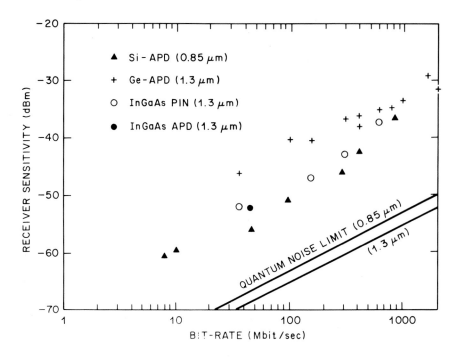

FIGURE 17. A comparison of the receiver sensitivity obtained using InGaAs p-i-n, InGaAs APD and Ge APD as the photodiodes. Comparison with that using Si APD at 0.85 μm is also shown. The solid lines indicate quantum noise limits for the 0.85 μm and 1.3 μm wavelength, respectively.

signals. The current gain used in the Si-APD was optimized for each bit-rate (ranging from about 20 at 1 Mb/sec to about 100 at 100 Mb/sec). The sensitivities for Si-APD receivers shown in Figure 17 represent about 200 signal photoelectrons per bit (averaged), a value which is only 12 dB from the theoretical quantum noise limit.

Because of the large dark current encountered in Ge and III—V alloy avalanche photodiodes, InGaAs PIN photodiodes and low-noise GaAs MESFET preamplifiers are used in present receivers for the 1.3 to 1.6 μm long wavelength region.[103-107] The PIN-FET receiver uses extremely low-input-capacitance (C < 0.5 pF) and high-transconductance (g_m ≥ 30 mS) GaAs FET transistors, along with low photodiode capacitance (C < 0.5 pF) and high input resistance (R ≥ 1 mΩ). Receiver sensitivities only 5 to 8 dB from those obtained with Si-APDs at 0.85 μm have been possible. However, the use of a large input resistor results in a large degree of equilization, and thus a lower dynamic range (≃20 dB), which has to be improved by other means.

Initial tests on a receiver at 45 Mb/sec using an InGaAs/InP APD indicated that the sensitivity was −53.2 dBm[84] which was 3 to 5 dB higher than the sensitivity of a PIN-FET receiver, and was only 2 to 3 dB lower than that of Si-APD at 0.85 μm wavelength.

Receiver sensitivities using Ge-APDs are comparable to PIN-FET receivers at bit-rates above several hundred Mb/sec, and are inferior at lower bit-rates.[107,108] The Ge-APD becomes more useful at very high bit-rates where the receiver noise is no longer dominated by the dark currents.

V. CONCLUSIONS

Silicon photodiodes with high sensitivity and sufficiently wide bandwidth for operation in the 0.8 to 0.9 μm wavelength region are now available commercially. Recently, long wavelength detectors based on InGaAs and InGaAsP PINs also have become available commercially, and are used in many experimental lightwave systems. The Ge avalanche diode still holds its place for high bit-rate applications until better InGaAs APDs become possible.

Current research is directed toward integration of the photodiodes with transistor amplifiers for low cost and high reliability. New materials such as the superlattice and new mechanisms such as resonance of split-off valence bands are being explored and are continuing to be of great interest in the search for low-noise, high sensitivity photodetector materials and structures. Much work lies ahead in this area of research.

REFERENCES

1. **Lee, T. P. and Li, Tingye,** Photodetectors, in *Optical Fiber Communications,* Miller, S. E. and Chynoweth, A. G., Eds., Academic Press, New York, 1979, chap. 18.
2. **Schinke, D. P., Smith, R. G., and Hartman, A. R.,** Photodetectors, in *Semiconductor Devices for Optical Communication,* Kressel, H., Ed., Springer-Verlag, 1980, chap. 3.
3. **Anderson, L. K. and McMurtry, B. J.,** High-speed photodetectors, *Proc. IEEE,* 54, 1335, 1966.
4. **Anderson, L. K., DiDomenico, J., Jr., and Fisher, M. B.,** High-speed photodetectors for microwave demodulation of light, *Adv. Microwaves,* 5, 1, 1970.
5. **Melchior, H.,** Demodulation and photodetection techniques, in *Laser Handbook,* Arrecchi, F. T. and Schulz-Dubois, E. D., Eds.,), Elsevier, Amsterdam, 1972, 725.
6. **Melchior, H.,** Sensitive high speed photodetectors for the demodulation of visible and near infrared light, *J. Lumin.,* 7, 390, 1973.
7. **Melchoir, H.,** Semiconductor detectors for optical communications, Conf. Laser Eng. Appl. 1973; *IEEE J. Quantum Electron.,* QE-9 (Abstr.), 659, 1973.
8. **Melchor, H.,** Detectors for lightwave communication, *Phys. Today,* 30, 32, 1977.
9. **Stillman, G. E. and Wolfe, C. M.,** Avalanche photodiodes, in *Semiconductors and Semimetals,* Vol. 12, Willardson, P. K. and Beers, A. C., Eds., Academic Press, New York, 1977, 291.
10. **Sze, S. M.,** *Physics of Semiconductor Devices,* John Wiley & Sons, New York, 1969, 59.
11. **Sze, S. M.,** *Physics of Semiconductor Devices,* John Wiley & Sons, New York, 1969, 27.
12. **McKay, K. G. and McAfee, K. B.,** Electron multiplication in silicon and germanium, *Phys. Rev.,* 91, 1079, 1953.
13. **Sze, S. M.,** *Physics of Semiconductor Devices,* John Wiley & Sons, New York, 1969, 61.
14. **Lee, C. A., Logan, R. A., Batdorf, R. L., Kleimack, J. J., and Weigman, W.,** Ionization rates of holes and electrons in silicon, *Phys. Rev.,* 134, A761, 1964.
15. **Miller, S. M.,** Avalanche breakdown in germanium, *Phys. Rev.,* 99, 1234, 1955.
16. **Logan, R. A. and Sze, S. M.,** Avalanche multiplication in Ge and GaAs p—n junctions, *J. Phys. Soc. Jpn. Suppl.,* 21, 434, 1966.
17. **Stillman, G. E., Wolfe, C. M., Foyte, A. G., and Lindley, W. T.,** Schottky barrier $In_xGa_{1-x}As$ alloy avalanche photodiodes for 1.06 μm, *Appl. Phys. Lett.,* 24, 8, 1974.
18. **Pearsall, T. P., Nahory, R. E., and Pollack, M. A.,** Impact ionization coefficients for electrons and holes in $In_{0.14}Ga_{0.86}As$, *Appl. Phys. Lett.,* 27, 330, 1975.
19. **Pearsall, T. P.,** Impact ionization rates for electrons and holes in $Ga_{0.47}In_{0.53}As$, *Appl. Phys. Lett.,* 36, 218, 1980.
20. **Pearsall, T. P., Nahory, R. E., and Pollack, M. A.,** Impact ionization rates for electrons and holes in $GaAs_{1-x}Sb_x$ alloy, *Appl. Phys. Lett.,* 28, 403, 1976.
21. **Melchior, H. and Lynch, W. T.,** Signal and noise response of high speed germanium avalanche photodiodes, *IEEE Trans. Electron Devices,* ED-13, 829, 1966.
22. **Goetzberger, A., McDonald, B., Haitz, R. H., and Scarlett, R. M.,** Avalanche effects in silicon p-n junctions. II. Structurally perfect junctions, *J. Appl. Phys.,* 34, 1591, 1963.
23. **Emmons, R. B.,** Avalanche photodiode frequency response, *J. Appl. Phys.,* 38, 1705, 1967.
24. **Kaneda, T., Takahashi, H., Matsumoto, H., and Yamaoka, T.,** Avalanche buildup time of silicon reach-through photodiodes, *J. Appl. Phys.,* 47, 4960, 1976.
25. **Goebloed, J. J.,** Comments on Avalanche buildup time of silicon reach-through photodiodes, *J. Appl. Phys.,* 48, 4004, 1977.
26. **McIntyre, R. J.,** Multiplication noise in uniform avalanche diodes, *IEEE Trans. Electron Devices,* ED-13, 164, 1966.

27. **McIntyre, R. J.,** The distribution of gains in uniformly multiplying avalanche photodiodes: theory, *IEEE Trans. Electron Devices,* ED-19, 703, 1972.

28. **Personick, S. D.,** New results on avalanche multiplication statistics with applications to optical detection, *Bell Syst. Tech. J.,* 50, 167, 1971.

29. **Personick, S. D.,** Statistics of a general class of avalanche detectors with applications to optical communication, *Bell Syst. Tech. J.,* 50, 3075, 1971.

30. **Conradi, J.,** The distribution of gains in uniformly multiplying avalanche photodiodes: experimental, *IEEE Trans. Electron. Devices,* ED-19, 713, 1972.

31. **Webb, P. P., McIntyre, R. J., and Conradi, J.,** Properties of avalanche photodiodes, *RCA Rev.,* 35, 234, 1974.

32. **Nishida, K.,** Avalanche-noise dependence on avalanche-photodiode structures, *Electron Lett.,* 13, 419, 1977.

33. **Ruegg, H. W.,** An optimized avalanche photodiode, *IEEE Trans. Electron Devices,* ED-14, 239, 1967.

34. **Conradi, J. and Webb, P. P.,** Silicon reach-through avalanche photodiodes for fiber optic applications, Proc. 1st Eur. Conf. Opt. Fire Commun., 1975, 128.

35. **Berchtold, K., Krumpholz, O., and Suri, J.,** Avalanche photodiodes with a gain-bandwidth product of more than 200 GHz, *Appl. Phys. Lett.,* 26, 585, 1975.

36. **Melchior, H., Hartman, A. R., Schinke, D. P., and Seidel, T. E.,** Planar epitaxial silicon avalanche photodiodes, *Bell Syst. Tech. J.,* 57, 1791, 1978.

37. **Kaneda, T., Matsumoto, H., and Yamaoka, T.,** A model for reach-through avalanche photodiodes (RAPD's), *J. Appl. Phys.,* 47, 3135, 1976.

38. **Kanbe, H., Kimura, T., Mizushima, Y., and Kajiyama, K.,** Silicon avalanche photodiodes with low multiplication noise and high-speed response, *IEEE Trans. Electron Devices,* ED-23, 1337, 1976.

39. **Nishida, K., Ishii, K., Minemura, K., and Taguchi, K.,** Double epitaxial silicon avalanche photodiodes for optical fiber communications, *Electron. Lett.,* 13, 280, 1977.

40. **Goebloed, J. J. and Smeets, E. T. J. M.,** Very low noise silicon planar avalanche photodiodes, *Electron. Lett.,* 14, 67, 1978.

41. **Conradi, J.,** Temperature effects in silicon avalanche diodes, *Solid-State Electron.,* 17, 99, 1974.

42. **Smith, R. G., Brackett, C. A., and Reinbold, H. W.,** Optical detector package, *Bell Syst. Tech. J.,* 57, 1809, 1978.

43. **Müller, J. and Ataman, A.,** Double-mesa thin-film reach-through silicon avalanche photodiodes with large gain bandwidth product, Tech. Dig., Int. Electron Devices Meet., Washington, D.C., December 6 to 8, 1976, 416.

44. **Takahashi, K., Takamiya, S., and Mitsui, S.,** A monolithic 1×10 array of silicon avalanche photodiodes, Tech. Dig., Int. Conf. Integr. Opt. Opt. Fiber Commun., Tokyo, July 1977, 37.

45. **Boyd, J. T. and Chen, C. L.,** Integrated optical silicon photodiode array, *Appl. Opt.,* 15, 1389, 1976.

46. **Pearsall, T. P., Nahory, R. E., and Cheilkowsky, J. R.,** Orientation dependence of free-carrier impact ionization in semiconductors: GaAs, *Phys. Rev. Lett.,* 39, 295, 1977.

47. **Capasso, F., Nahory, R. E., Pollack, M. A., and Pearsall, T. P.** Observation of electronic band structure effects on impact ionization by temperature tuning, *Phys. Rev. Lett.,* 39, 723, 1977.

48. **Hidebrand, O., Kuebart, W., Benz, K. W., and Pilkuhn, M. H.,** $Ga_{1-x}Al_xSb$ avalanche photodiodes: resonant impact ionization with very high ratio of ionization coefficients, *IEEE J. Quan. Electron.,* QE-17, 284, 1981.

49. **Riesz, R. P.,** High speed semiconductor photodiodes, *Rev. Sci. Instrum.,* 33, 994, 1962.

50. **Conradi, J.,** Planar germanium photodiodes, *Appl. Opt.,* 14, 1948, 1975.

51. **Ando, J., Kanbe, H., Kimura, T., Yamaoka, T., and Kaneda, T.,** Characteristics of germanium avalanche photodiodes in the wavelength region of 1.0-1.6 μm, *IEEE J. Quantum Electron.,* QE-14, 804, 1978.

52. **Mikawa, T., Kagawa, S., Kaneda, T., Sakurai, T., Ando, J., and Mikami, O.,** A low-noise n^+np Germanium Avalanche Photodiode, *IEEE J. Quantum Electron.,* QE-17, 210, 1981.

53. **Kanbe, H., Grosskopf, G., Mikami, O., and Machida, S.,** Dark current noise characteristics and temperature dependence in germanium avalanche photodiodes, *IEEE J. Quantum Electron.,* QE-17, 1534, 1981.

54. **Smith, R. G. and Personick, S. D.,** Receiver design for optical fiber communication systems, in *Semiconductor Devices for Optical Communication,* Kressel, H., Ed., Springer-Verlag, Berlin, 1979, chap. 4.

55. **Pearsall, T. P. and Papuchon, M.,** The $Ga_{0.47}In_{0.53}As$ homojunction photodiode — a new avalanche photodetector in the near infrared between 1.0 and 1.6 μm, *Appl. Phys. Lett.,* 33, 640, 1978.

56. **Washington, M. A., Nahory, R. E., and Beebe, E. D.,** High-efficiency $In_{1-x}Ga_x$—As_yP_{1-y}/InP photodetectors with selective wavelength response between 0.9 nd 1.7 μm, *Appl. Phys. Lett.,* 33, 854, 1978.

57. **Hurwitz, C. E. and Hsieh, J. J.,** GaInAsP/InP avalanche photodiodes, *Appl. Phys. Lett.,* 32, 487, 1978.

58. **Olsen, G. H. and Kressel, H.,** Vapor-grown 1.3 μm InGaAsP/InP avalanche photodiodes, *Elect. Lett.,* 15, 141, 1979.

59. **Lee, T. P., Burrus, C. A., and Dentai, A. G.,** InGaAsP/InP photodiodes: microplasma-limited avalanche multiplication at 1–1.3 μm wavelength, *IEEE J. Quantum Electron.,* QE-15, 30, 1979.

60. **Burrus, C. A., Dentai, A. G., and Lee, T. P.,** InGaAsP p-i-n photodiodes with low dark current and small capacitance, *Electron Lett.,* 15, 655, 1979.

61. **Lee, T. P., Burrus, C. A., and Dentai, A. G.,** InGaAs/InP p-i-n photodiodes for lightwave communication at 0.95 to 1.65 μm wavelength, *IEEE J. Quant. Electron.,* QE-17, 232, 1981.

62. **Forrest, S. R., DiDomenico, M., Jr., Smith, R. E., and Stocker, H. J.,** Evidence of tunneling in reverse-biased III-V photodetector diodes, *Appl. Phys. Lett.,* 36, 580, 1980.

63. **Forrest, S. R.,** Performance of $In_xGa_{1-x}As_yP_{1-y}$ photodiodes with dark current limited by diffusion, generation recombination, and tunneling, *IEEE J. Quant. Electron,* QE-17, 217, 1981.

64. **Ando, H., Kanbe, H., Ito, M., and Kaneda, T.,** Tunneling current in InGaAs and optimum design for InGaAs/InP avalanche photodiode, *Jpn. J. Appl. Phys.,* 19, L277, 1980.

65. **Lee, T. P., Burrus, C. A., Ogawa, K., and Dentai, A. G.,** Very high speed back-illuminated InGaAs/InP p-i-n punch-through photodiodes, *Electron. Lett.,* 17, 431, 1981.

66. **Lee, T. P., Burrus, C. A., Cho, A. Y., and Cheng, K. Y.,** Zn-diffused back-illuminated p-i-n photodiodes in InGaAs/InP grown by molecular beam epitaxy, *Appl. Phys. Lett.,* 37, 730, 1980.

67. **Susa, N., Yamauchi, Y., and Kanbe, H.,** Planar photodiodes made from vapor-phase epitaxial $In_xGa_{1-x}As$, *Electron. Lett.,* 15, 238, 1979.

68. **Susa, N., Yamauchi, Y., Ando, H., and Kanbe, H.,** *IEEE Electron Dev. Lett.,* EDL-1, 55, 1980.

69. **Forrest, S. R., Camlibel, I., Kim, O. K., Stocker, H. J., and Zuber, J. R.,** Low dark-current, high-efficiency planar $In_{0.53}Ga_{0.47}As$/InP PIN photodiodes, *IEEE Elect. Lett.,* EDL-2, 283, 1981.

70. **Diadiuk, V., Groves, S. H., and Hurwitz, C. E.,** Avalanche multiplication and noise characteristics of low dark current GaInAsP/InP avalanche photodiodes, *Appl. Phys. Lett.,* 37, 807, 1980.

71. **Campbell, J. C., Lee, T. P., Dentai, A. G., and Burrus, C. A.,** Dual-wavelength demultiplexing InGaAsP photodiode, *Appl. Phys. Lett.,* 34, 401, 1979.

72. **Lee, T. P., Campbell, J., C., Ogawa, K., McCormick, A. R., Dentai, A. G., and Burrus, C. A.,** Dual-wavelength 1.5 Mb/s lightwave receiver employing an InGaAsP wavelength demultiplexing detector, *Electron. Lett.,* 15, 388, 1979.

73. **Campbell, J. C., Dentai, A. G., Lee, T. P., and Burrus, C. A.,** Improved two-wavelength demultiplexing InGaAsP photodetectors, *IEEE J. Quant. Electron.,* QE-16, 601, 1980.

74. **Ogawa, K., Lee, T. P., Burrus, C. A., Campbell, J. C., and Dentai, A. G.,** Wavelength division multiplexing experiment employing dual-wavelength LED and photodiodes, *Electron Lett.,* 17, 857, 1981.

75. **Nishida, K., Taguchi, K., and Matsumoto, Y.,** InGaAsP heterostructure avalanche photodiodes with high avalanche gain, *Appl. Phys. Lett.,* 35, 251, 1979.

76. **Kanbe, H., Susa, N., Nakagome, H., and Ando, H.,** InGaAs Avalanche photodiode with InP p-n junction, *Electron. Lett.,* 16, 163, 1980.

77. **Susa, N., Nakagome, H., Mikami, O., Ando, H., and Kanbe, H.,** New InGaAs/InP avalanche photodiode structure for 1-1.6 μm wavelength region, *IEEE J. Quant. Electron.,* QE-16, 864, 1980.

78. **Susa, N., Nakagome, H., Ando, H., and Kanbe, H.,** Characteristics in InGaAs/InP avalanche photodiodes with separated absorption and multiplication regions, *IEEE J. Quant. Electron.,* QE-17, 243, 1981.

79. **Kim, O. K., Forrest, S. R., Bonner, W. A., and Smith, R. G.,** A high gain $In_{0.53}Ga_{0.47}As$/InP avalanche photodiode wih no tunneling leakage current, *Appl. Phys. Lett.,* 39, 402, 1981.

80. **Matsushima, Y., Sakai, K., and Noda, Y.,** New type InGaAs/InP heterostructure avalanche photodiode with buffer layer, *IEEE Electron Device Lett.,* EDL-2, 179, 1981.

81. **Yeats, R. and Von Dessonneck, K.,** Detailed performance characterstics of hybrid InP-InGaAsP APDs, *IEEE Electron Device Lett.,* EDL-2, 268, 1981.

82. **Shirai, T., Osaka, F., Yamasaki, S., Nakajima, K., and Kaneda, T.,** 1.3 μm InP/InGaAsP planar avalanche photodiodides, *Electron Lett.,* 17, 826, 1981.

83. **Diadiuk, V., Groves, H., Hurwitz, C. E., and Iseler, G. W.,** Low dark current, high-gain GaInAs/InP avalanche photodiodes, *IEEE J. Quantum Electron,* QE-17, 260, 1981.

84. **Forrest, S. R., Williams, G. F., Kim, O. K., and Smith, R. G.,** Excess-noise and receiver sensitivity measurement of $In_{0.53}Ga_{0.47}As$/InP avalanche photodiodes, *Electron. Lett.,* 17, 917, 1981.

85. **Law, H. D., Tomasetta, L. R., Nakano, K., and Harris, J. S.,** 1.0 μm to 1.4 μm high speed avalanche photodiode, *Appl. Phys. Lett.,* 33, 416, 1978.

86. **Capasso, F., Panish, M. B., Sumski, S., and Foy, P. W.,** Very high quantum efficiency GaSb mesa photodetectors between 1.3 and 1.6 μm, *Appl. Phys. Lett.,* 36, 165, 1980.

87. **Capasso, F., Panish, M. B., and Sumski, S.,** The liquid phase epitaxial growth of low net donor concentration ($5 \times 10^{14} - 5 \times 10^5$/cm³) GaSb for detector applications in the 1.3—1.6 μm region, *IEEE J. Quantum Electron.,* QE-17, 273, 1981.

88. **Law, H. D., Chin, R., Nakano, K., and Milano, R. A.,** The GaAlAsSb quaternary and GaAlSb ternary alloys and their application to infrared detectors, *IEEE J. Quantum Electron.,* QE-17, 275, 1981.

89. **Beneking, H., Mischel, P., and Schul, G.,** High-gain wide-gap-emitter, $Ga_{1-x}Al_xAs$-GaAs phototransistor, *Electron. Lett.,* 12, 395, 1976.

90. **Milano, R. A., Windhorn, T. H., Anderson, E. R., Stillman, G. E., Dupuis, R. D., and Dapkus, P. D.,** $Al_{0.5}Ga_{0.5}As$-GaAs heterojunction photo-transistors grown by metalorganic chemical vapor deposition, *Appl. Phys. Lett.,* 34, 562, 1979.

91. **Wright, P. D., Nelson, R. J., and Cella, T.,** High-gain InGaAsP-InP heterojunction photo-transistors, *Appl. Phys. Lett.,* 37 192, 1980.

92. **Alavi, K. T., Markunus, R. J., and Fonstad, C. C.,** LPE-grown InGaAsP/InP heterojunction bipolar photo-transistors, Tech. Dig., Int. Electron Devices Meeting, Washington, D.C., December 3 to 5, 1979, 643.

93. **Campbell, J. C., Dentai, A. G., Burrus, C. A., and Furguson, J. F.,** InP/InGaAs heterojunction photo-transistors, *IEEE J. Quantum Electron.,* QE-17, 264, 1981.

94. **Gammel, J. C., Ohno, H., and Ballantyne, J. M.,** High-speed photoconductive detectors using GaInAs, *IEEE J. Quantum Electron.,* QE-17, 269, 1981.

95. **Gammel, J. C., Metze, G. M., and Ballantyne, J. M.,** A photoconductive detector for high-speed fiber communication, *IEEE Trans. Electron Devices,* ED-28, 841, 1981.

96. **Barnard, J., Ohno, H., Wood, C. E. C., and Eastman, L. F.,** Integrated double-heterostructure $Ga_{0.47}In_{0.53}As$ photoreceiver with automatic gain control, *IEEE Electron Device Lett.,* EDL-2, 7, 1981.

97. **Copeland, J. C., Dentai, A. G., and Lee, T. P.,** p-n-p-n Optical detectors and light-emitting diodes, *IEEE J. Quantum Electron.,* QE-14, 810, 1978.

98. **Leheny, R. F., Nahory, R. E., Pollack, M. A., Ballman, A. A., Beebe, F. D., DeWinter, J. C., and Martin, R. J.,** Integrated $In_{0.53}Ga_{0.47}As$ p-i-n FET photoreceiver, *Electron. Lett.,* 16, 353, 1980.

99. **Chin, R., Holonyak, N., Jr., Stillman, G. E., Tang, J. Y., and Hess, K.,** Impact ionization in multilayered heterojunction structures, *Electron Lett.,* 16, 467, 1980.

100. **Capasso, F., Tsang, W. T., Hutchinson, A. L., and Williams, G. F.,** A novel multilayer avalanche photodetector for low-noise application, Digest, Integrated and Guided wave Optics Meeting, January 1982, ThC2; also, enhancement of electron impact ionization in a super lattice. A new avalanche photodiode with a large ionization rate ratio, *Appl. Phys. Lett.,* 40, 38, 1982.

101. **Personick, S. D.,** Receiver design for digital fiber optic communication systems. Parts I and II, *Bell Syst. Tech. J.,* 52, 843, 1973.

102. **Personick, S. D.,** Receiver design, in *Optical Fiber Communications,* Miller, S. E. and Chynoweth, A. G., Eds., Academic Press, New York, 1979, chap. 19.

103. **Smith, D. R., Hooper, R. C., and Garrett, I.,** Receivers for optical communications: a comparison of avalanche photodiodes with PIN-FET Hybrids, *Opt. Quantum Electron.,* 10, 293, 1978.

104. **Lee, T. P., Burrus, C. A., Dentai, A. G., Ogawa, K., and Chinnock, E.,** InGaAs PIN photodetector and GaAs FET amplifiers used at 1.3 μm wavelength in 45 Mb/s and 274 Mb/s experimental repeaters, *Tech. Dig. CLEOS'80,* San Diego, February 26 to 28, 1980, 26.

105. **Lee, T. P., Burrus, C. A., Dentai, A. G., and Ogawa, K.,** Small-area InGaAs/InP p-i-n photodiodes: fabrication, characteristics and performance of devices in 274 Mb/s and 45 Mb/s lightwave receivers at 1.31 μm wavelength, *Electron. Lett.,* 16, 155, 1980.

106. **Smith, D. R., Hooper, R. C., Ahmad, K., Jenkin, D., Mabbitt, A. W., and Nicklin, R.,** p-i-n/FET hybrid optical receiver for longer-wavelength optical communication systems, *Electron. Lett.,* 16, 69, 1980.

107. **Smith, D. R., Hooper, R. C., Smyth, P. P., and Rejman, M. A.,** Experimental comparison of two long wavelength optical receivers, Digest Optical Fiber Conference, Phoenix, Arizona, April 1982, 16.

108. **Kimura, T.,** Single-mode digital transmission technology, *Proc. IEEE,* 68, 1263, 1980.

Chapter 6

DIGITAL OPTICAL COMMUNICATION SYSTEMS

E. E. Basch and H. A. Carnes

TABLE OF CONTENTS

I. INTRODUCTION

The preceding chapters have summarized the relevant characteristics of the sources, detectors, and fibers used in designing a fiber optic communication system and clarified some of the trade-offs that need to be made in selecting the proper component when designing a system. In this chapter we will provide a mainly heuristic approach to designing fiber optic communications systems and avoid sophisticated mathematical tools and derivations of analytical results seldom used by most practicing telecommunications engineers. Instead, the reader will be directed to relevant references for in-depth analysis of the subject areas.

Over the past 15 years, optical fiber communication has evolved from a laboratory curiosity to a commercial reality. Today, there is a proliferation of system experiments, economic studies, and hardware appearing for sale. These commercial systems primarily transmit digital signals with moderate (200 Mb/sec) bit rates through graded-index multimode fibers and operate in the 850 nm region. The application of analog transmission (multichannel voice and CATV) has remained small, primarily because optical communication is more advantageous when digital transmission is used and because of the accelerating conversion from analog to digital technology. In an effort to extend the repeater spacing of long haul trunks, systems operating in the long wavelength region of 1.3 μm are being introduced commercially.[1,2] Many of the fundamental issues in the analysis and design of multimode optical communications systems have not been investigated in detail. As a case in point, the exceptionally long repeater spacings made possible by the very low attenuation of fibers in the long wavelength region is putting very stringent requirements on fiber bandwidth, but adaptive equalization[3] and multilevel coding techniques[4] have largely been neglected.

Although, from a requirement point of view, many short haul telecommunication applications can be addressed adequately by multimode technology, research and development activities are shifting to single-mode technology.[5] The use of single-mode fibers permits one to realize simultaneously all the advantages of maximized bandwidth and minimized loss. Exceptionally long repeater spacings, even at high bit rates, are clearly one possibility; alternately, provided a substantial power margin has been designed into the links, the same fiber will be capable of upgrading to higher bit rate operation as terminal technology advances.

A more speculative approach involving single-mode fibers is the use of coherent optical transmission[6] in which a heterodyne-type detection process is used. Heterodyne-type systems have been extensively studied for free space optical communications,[7] and their feasibility in conjunction with optical fibers is currently under investigation.[8-10]

II. SYSTEM DESIGN

This section will focus on the general principles of the design of a digital fiber optic link. These principles will apply to any digital communication link, regardless of its specific application. In-depth discussion of the transmitter, receiver, and fiber subsystems will be covered in subsequent sections.

A distinct characteristic of digital transmission systems is that, in a cascade of repeated links, each individual link may be analyzed alone. This is true of noise and bandwidth; however, clock jitter may at least in part be cumulative. If one ignores the cumulative clock jitter problem, the worst case individual link sets the performance level of the entire link. The remaining discussion will consider the jitter free case, since the subject of timing recovery and its associated jitter have been treated thoroughly in the literature.[11-13]

The performance of a digital transmission system can be conveniently quantified by its bit rate distance product, i.e., the transmitter-receiver spacing that is feasible at the desired bit rate. Many parameters influence this system performance; foremost are the available transmitter power, the required input power at the receiver to obtain the desired bit error rate (BER), and the overall system loss and bandwidth. The above numbers should, of course, take into account normal aging and environmental changes.

The received optical power is naturally related to the power launched into the fiber waveguide by the transmitter and the losses encountered as the optical signal passes through fibers, connectors, and splices. Bandwidth is determined primarily by the fiber and the type of optical source. Figure 1 shows a typical transmission system block diagram consisting of a transmitter, receiver, and several sections of fiber. The loss of each fiber section will be designated as ℓ_N and each splice or connector loss as S_N. The difference between the transmitter power and the receiver sensitivity (at the desired BER) defines the optical link power budget. In order to maintain the specified BER, the insertion loss between the optical source and detector must not exceed the power budget value. The relationship between the power budget and link insertion loss can be expressed by the following relationship based on the model of Figure 1.

$$\overline{P}_S - \overline{P}_{R_{min}} > \Sigma \, S_N + \Sigma \, \ell_N + k_E \tag{1}$$

where $\overline{P}_S$ is the average power available at the optical source's fiber flylead in dBm; $\overline{P}_R$ is the receiver sensitivity in dBm; S_N is the splice loss in dB; ℓ_N is the loss of a section of fiber in dB; and K_E is the excess loss due to a nonequilibrium mode distribution.

Launch conditions at the interface between the optical source and fiber waveguide affect the actual end-to-end insertion loss of a fiber link. Light emitting diodes are incoherent sources that tend to launch so-called leaky modes and cladding modes into the fiber that may result in a substantial excess loss, K_E, over the nominal loss of the fiber. Excess losses of 3 to 4 dB are not uncommon for surface-emitter LEDs, but are usually small or even negligible for edge-emitter LEDs and injection laser diodes. This excess loss usually takes place within meters to hundreds of meters from the source and should be determined experimentally before a power budget analysis takes place.

It is generally good practice to include a safety factor referred to as the optical power margin in the optical power budget. The margin is the difference between the nominal receiver sensitivity and the higher value of the optical power received. A margin in received power can be used to maintain the system BER in the presence of noise phenomena that are known to occur but difficult to predict quantitatively. Systems using injection laser diodes, for example, are susceptible to modal noise and external cavity reflections that will degrade system performance if insufficient margins exist.

A power budget analysis will determine whether or not a system is loss-limited, that is, if the received power is sufficient to meet or exceed a BER specification. However, a power budget analysis is not a sufficient condition to guarantee a minimum BER performance. System performance can also be delay distortion limited (often, but less accurately referred to as dispersion limited systems). In such a system, the maximum spacing between transmitter and receiver is limited by (excessive) pulse broadening in the fiber, which causes intersymbol interference with the attendant reduction in BER performance. Effectively, the pulse spreading has lowered the receiver sensitivity and thus the optical link power budget. Eye patterns[14] provide a quick way to check if intersymbol interference is present; Figure 2 shows an eye pattern of a binary signal with eye closure due to intersymbol interference. The delay distortion in the fiber is combined with pulse broadening in the transmitter and receiver to cause the overall intersymbol interference in the system. There are two sources of delay distortion in fibers: modal distortion and chromatic dispersion. Modal distortion (often referred to as modal dispersion) results from differential mode delay. Chromatic dispersion describes the process by which an electromagnetic signal is distorted because the various wavelength components of that signal have different propagation characteristics. In principle, the distorted pulses can be electrically equalized.[15,16] However, equalization will enhance

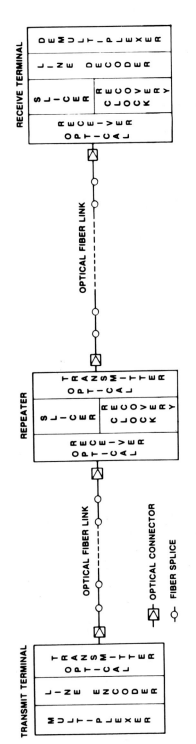

FIGURE 1.   Optical transmission system block diagram.

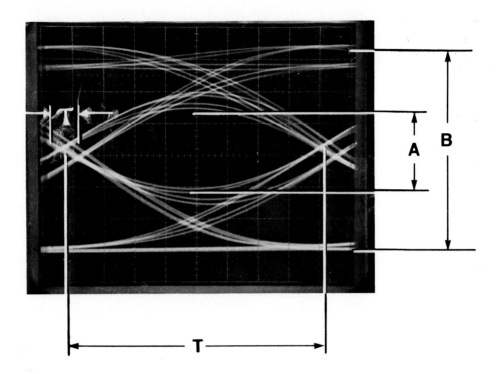

%EYE CLOSURE = A/B × 100%
%JITTER = τ/T × 100%

FIGURE 2.   Bandlimited eye pattern.

various noise components in the received pulses and, therefore, causes a reduction in receiver sensitivity. The question of how to design a practical equalizing receiver for optical transmission systems has not been fully investigated. It has been established by analysis[17,18] that if the product of the bit rate and the root mean square (rms) width of the impulse response of the system (up to and including the detector) is less than 0.25, the intersymbol interference has a negligible effect (i.e., $1 <$ dB) on the receiver sensitivity.*

It is often more convenient to analyze the rise time in a system than to determine the rms impulse response. As a rough rule of thumb, for intersymbol interference to have a negligible influence on receiver sensitivity in a digital transmission link, the pulse rise time of the system should not exceed 45% (Gaussian waveform) to 70% (exponential waveform) of the bit period for nonreturn-to-zero (NRZ) pulses.[19] In systems in which the system rise time exceeds 70% of the bit period, the receiver sensitivity penalty can be a strong function of the received pulse shape as well as the specific receiver design. In order to calculate the overall system impulse response or end-to-end rise time, one combines the effects of individual contributors as the square root of the sum of the squares,

$$\sigma_{out}^2 = \Sigma \sigma_n^2 \qquad (2)$$

where σ_{out} is rms width of the output pulse and σ_n the RMS width of the impulse response

---

* These rules assume a receiver with Nyquist filtering. Receivers that do not provide a Nyquist response usually exhibit a somewhat larger reduction in sensitivity.

of the nth contributor. For chromatic dispersion, σ increases linearly with fiber length. Pulse spreading due to differential mode delay can increase less than linearly in fiber length due to mode mixing effects. Since this mode mixing is uncontrolled, it is not a fully reliable system parameter.

For systems with a Gaussian pulse response, one can derive simple relations between rise time, rms pulse width, full width, half maximum pulse width, and the 3 dB electrical and optical bandwidth. These are

$$\text{BW(3 dB optical)} \approx \frac{0.187}{\sigma_{rms}} \text{ Hz} \tag{3}$$

$$\text{BW(3 dB electrical)} \approx \frac{0.133}{\sigma_{rms}} \text{ Hz} \tag{4}$$

$$T_R \text{ (rise time)} = \frac{0.317}{\text{BW(3 dB optical)}} \text{ sec*} \tag{5}$$

$$\sigma_{rms} = \frac{\sigma_{FWHM}}{2.36} \tag{6}$$

This section has dealt with the techniques to establish the feasibility of a digital fiber optic transmission link. A set of parameters, which includes optical source power, fiber attenuation, fiber bandwidth, and receiver sensitivity, are known or can be easily calculated. The system requirements are BER value, link distance, and bit rate. With the known parameters and some simple calculations, the number of repeater fiber links may be determined. There are, however, a large number of options as to the componentry available for a particular transmission application. Certainly the most critical decision is the choice of operating wavelength. Three wavelength regions are available to the system designer, the 850 nm region, the 1300 nm region, and the 1550 nm region. Each region, in sequential order, gives progressively lower attenuation. Next in the line of options is the index of refraction profile of the fiber waveguide. The selection of wavelength and fiber profile is quite dependent on the system bit rate and the distance the information is to be sent. Secondary decisions must be made as to the choice of optical source and detector. The remaining sections will investigate in more detail the subsystems of a digital fiber optic link.

III. OPTICAL TRANSMITTER

The leading light sources for fiber optic communications are the light emitting diode and injection laser diode. Their compatibility with the spectral transmission characteristics and physical dimensions of optical fibers and reasonable lifetimes have made possible the rapid expansion of optical fiber communications. Optical sources used in communication systems are capable of launching tens of microwatts to several milliwatts of optical power into fibers with 0.2 to 0.5 numerical apertures (NA). Optical diodes are also quite easily modulated with straight forward electronic circuitry for low to moderate bit rate systems. The characteristics of optical light sources have been described in detail in Chapter 4. Here, we will focus on the operations and performance parameters important to digital optical transmitters.

In order to transport information from the transmitter to the receiver, it is necessary for the information to modulate the amplitude, frequency, or phase of the optical signal.[20] Frequency and phase modulation would require extremely stable optical sources and a heterodyne detection technique, which is not a practical approach at present. There are two

---

* For exponential wave forms $T_R = 0.35/\text{BW (3 dB electrical)}$.

methods used to modulate the amplitude of the optical signal. The first is direct modulation in which the source output power is varied directly by modulating the bias current of the device. In external modulation, the second approach, the bias current through the optical source is held constant and a separate external modulator varies the amount of optical power that is coupled into the fiber waveguide. External modulators[20,21] are usually based upon electrooptic, magnetooptic, or acoustooptic effect. Direct modulation is the more commonly used and practical method at present.

In many applications, the system designer must take into account the spectral width of the source as well as its total power output. Current optical sources have relatively wide spectral width, 30 to 60 nm for LEDs and 1 to 5 nm for injection laser diodes. By comparison, a modulation bandwidth of 100 GHz would require less than 1 nm in spectral width. Chromatic dispersion in the optical fiber causes the different wavelengths emitted by the optical source to travel at different velocities in the fiber. As pointed out in the previous section, that is one of the causes of pulse broadening in optical fibers. Another phenomenon occurs in laser diodes where the distribution of power among the various spectral components (longitudinal modes) can vary significantly relative to total power output, both within a pulse, and from pulse to pulse. Here the chromatic dispersion in the fiber causes an amplitude variation of the signal at the receiver output, called mode partition[22-25] noise. The signal-to-noise ratio due to mode partition noise is independent of signal power, so that the overall system error rate cannot be improved beyond the limit set by this noise. This is an important difference with regard to receiver sensitivity degradation normally associated with chromatic dispersion, which can be compensated for by increasing the signal power.

Chromatic dispersion decreases with increasing wavelength, until it reaches a minimum around 1300 nm, after which it increases again. We will treat chromatic dispersion in optical fibers in more detail in the section on fiber subsystems. Since the pulse broadening due to chromatic dispersion is proportional to the spectral width of the source, it can be reduced by limiting the spectral width of the source. With an LED, this can be accomplished by placing an optical filter between source and detector. Inevitably, such a filter will have some insertion loss, and trade-offs must be made between the additional loss and the beneficial effects of reducing chromatic dispersion.[26,27] In laser diodes the spectral width has been reduced through the use of shorter than "normal" cavity length[28] and also by adding an external optical cavity.[29]

The modulation bandwidth of LEDs and laser diodes depends on the electrical characteristics of the driving circuit and the carrier recombination lifetimes. The latter have been discussed in more detail in Chapter 4, Section III.D. Figure 3 shows circuitry capable of switching a high radiance surface emitting LED in 4 nsec. The modulating signal is applied to a differential amplifier formed by the transistor pair of Q1-1 and Q1-2. This stage guarantees that the LED current pulse is uniform in amplitude and switching characteristics. A current source, indicated as transistor Q1-4, controls the differential amplifier gain and the LED bias current amplitude. The differential amplifier output supplies an in phase and reverse phase signal to the base of the output push-pull stage of transistors Q3 and Q4. The base drive signal to Q4 is first level shifted by transistor stages Q2 and Q1-3. Figure 4 shows the LED bias current and Figure 5 the resultant optical pulse out of a short fiber fly-lead with a 0.2 NA. Note in Figure 4 the 30% overshoot in the bias current waveform. Current peaking is accomplished by a high frequency preemphasis network R_{11} and C_4 (light on) and R_{16} and C_7 (light off). The current peaking bias causes a rapid in-rush of carriers into the active region that rapidly changes the LED depletion layer capacitance. The peaked current pulse approximately doubles the switching speed of the optical pulse. The particular case shown in Figure 5 has a 4 nsec rise and fall time. The combined effect of material dispersion (for GaAlAs LEDs) and switching speed capability limits the use of LED devices to systems with bit rates between 50 Mb/sec and 100 Mb/sec. Heavier doping in the recom-

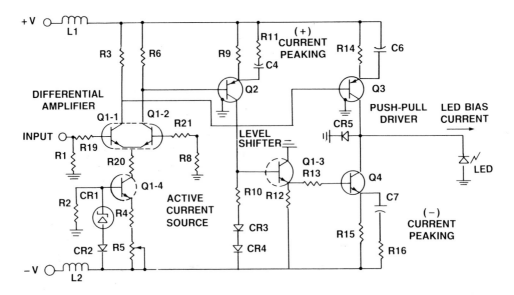

FIGURE 3. High speed LED modulator.

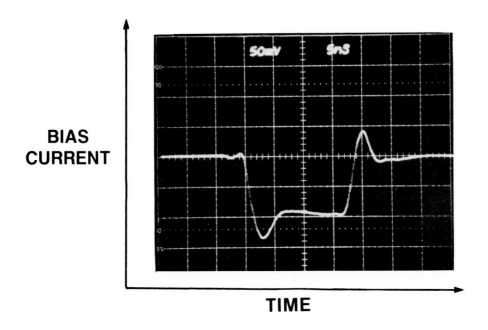

FIGURE 4. LED drive current generated by circuit shown in Figure 3.

bination region will increase switching speed at the expense of quantum efficiency. However, with operation at wavelengths in the 800 to 900 nm region, dispersion would limit the optical transmission distance to 2 to 3 km. When considering operation in the 1300 nm region, the limits imposed by fiber material dispersion for all practical purposes are eliminated.

The optical power generated by a LED must be coupled into a fiber to be useful for communication purposes. Parameters that determine the efficiency of the coupling are the ratio of the fiber core area to the active area of the LED, the fiber NA, and the far-field emission pattern of the light source. Far-field characteristics of light sources were discussed in detail in Chapter 4, Section III.C. If the fiber core is significantly larger than the active

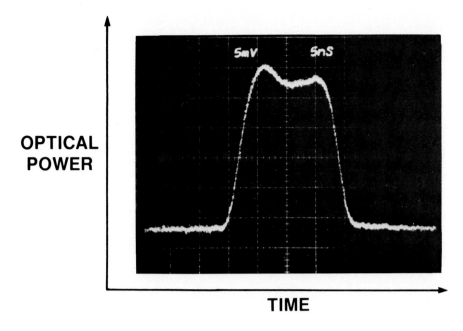

OPTICAL
POWER

TIME

FIGURE 5. Optical pulse generated by drive current shown in Figure 4.

area of the source, focusing optics can be used to increase the coupling efficiency. Without such optics, the coupling efficiency for surface emitting LEDs is proportional to the square of the fiber NA. Thus, one desires fibers with a large NA to increase the coupling efficiency between the LED and the fiber. However, the pulse spreading associated with modal distortion also increases proportional to the square of the fiber NA. Typically, NAs for fibers used in communication links range from 0.2 to 0.5. Short haul, low bit rate applications generally use a large NA and large core diameter fiber. Long haul, moderate to high bit rate systems use the smaller core, lower NA fiber. Figure 6 shows the typical light output of a high radiance, surface emitting LED coupled to a 50-μm core-graded index fiber. The characteristic is fairly linear at first, then saturates gradually due to self-heating effects. Temperature influences the output power of an LED through its effects on the internal quantum efficiency of the device, which decreases exponentially with increasing temperature. A temperature rise also causes the output spectrum to broaden and a shift in the peak wavelength of a few angstroms/°C. Long wavelength (1300 to 1500 nm), InGaAsP LEDs exhibit a higher temperature dependence than the shorter wavelength (800 to 900 nm) GaAs devices due to characteristic material difference. For example, a typical long wavelength device will experience more than 2 dB decrease in output power when the temperature increases from 20 to 50°C. Over the same temperature range, the output power of a typical short wavelength LED will decrease less than 1 dB. Usually, no attempt is made in LED transmitters to stabilize the operating temperature or to compensate for temperature induced changes in the output power.

Systems employing LEDs as light sources are characterized by a relatively low bit rate distance product. Chromatic dispersion in the fiber forms an important limitation for diodes operating in the 800- to 900-nm wavelength region. Longer wavelength LEDs can be used in systems with significantly larger bit rate distance products because of the reduction in dispersion and lower attenuation of the fiber.

The other light source of interest is the injection laser diode (ILD). Laser diode sources are capable of coupling more power more efficiently into an optical fiber. Furthermore, they can be directly modulated at rates up to several gigahertz per second and have a much

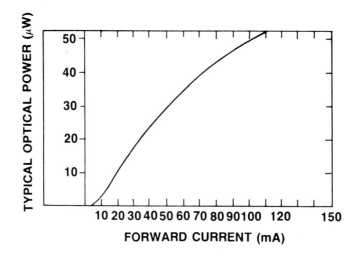

FIGURE 6.   Typical optical power emitted from a LED with a 2m length of 0.2 NA fiber, as a function of input current.

narrower spectral width than the LED. Thus laser diodes are used in systems with moderate to high bit rates and/or long repeater spacings. Disadvantages of laser diodes are more complex transmitter circuitry and three sources of noise (modal noise,[30,31] mode partition noise,[22-25] optical feedback noise[32]) created by interactions between laser and fiber.

Figure 7 shows the light output vs. current characteristic of a typical AlGaAs double heterostructure (discussed in Chapter 4) laser diode operating at 840 nm. The laser output characteristic has two distinct regions. The low output power region is characterized by spontaneous emission with very little power coupled into the fiber. In this region, the laser diode behaves as an LED. Once a sufficiently high current is reached, referred to as the threshold current, lasing or stimulated emission occurs. Beyond the threshold current point, the laser diode has a narrow spectral width, complete spatial coherence (assuming single transverse mode operation), and the fiber coupled quantum efficiency of the device increases dramatically so that only a small increase in drive current is needed to generate several milliwatts of optical power.

Temperature variations are much more of a concern with injection laser diodes, ILD, than with LEDs. Figure 7 shows the temperature dependence of the power-current characteristics of an AlGaAs laser. The threshold current increases as an exponential function of junction temperature.

$$I = I_{TH}e^{-T/T_o} \tag{7}$$

where T is the junction temperature (°K) and To is an empirical constant (°K). The parameter T_o is generally between 100 and 150°K for AlGaAs, short wavelength lasers, and between 60 and 70 °K for InGaAsP, long wavelength lasers. Maintaining a constant output power with an ILD in a changing temperature environment requires feedback control of either temperature or output power. Threshold current also increases with age of the laser. This is a gradual process in a "healthy" ILD but must still be considered in the design of a modulator circuit. The safest approach is to include a feedback control for the output optical power and a temperature regulator. The use of a temperature regulator has a secondary benefit in that it can extend the lifetime of the laser. Solid state laser lifetime is very dependent on the junction temperature. Operating at temperatures substantially above room temperature greatly accelerates the device aging process.

Temperature stabilization of the junction is easily achieved with a Peltier effect thermo-

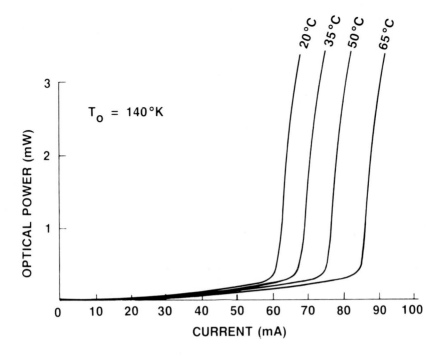

FIGURE 7.   Output characteristic of a typical stripe geometry double heterostructure AlGaAs, ILD with temperature as a parameter.

electric cooler. The ILD is mounted on one surface of the thermoelectric cooler along with a thermister. Temperature is sensed by the thermister, which provides an error signal for a feedback loop. The output of the feedback loop terminates in a controlled current source that drives the thermoelectric cooler.

An alternative stabilization technique is to allow the ILD junction temperature to change and compensate by varying the DC bias current. A photodetector is mounted in front of the ILD rear facet. The current generated by the photodetector is proportional to the optical power launched into the fiber waveguide. Figure 8 shows a circuit diagram that both pulses the ILD optical power and maintains a constant output average power. The photodetector current is compared to a reference related to the desired average power. As the threshold current varies with temperature or age, the photodetector senses the change. An error signal is generated and integrated by a pair of operational amplifiers. A controlled current source modifies the DC ILD bias based on the integrated error signal. Caution should be taken with an averaging type controller as its performance is affected by the modulation duty cycle. The control loop discussed above requires the modulation to have an equal probability of a mark or space. Long sequences of NRZ marks, exceeding the time constant of the feedback loop, will result in the optical mark pulse to decay to the average power set by the reference level. Alternative strategies are possible that require high speed control of the mark and space power levels.

Modulation of the optical power is accomplished by combining the bias current with the signal current. This results in two optical intensity levels representing the electrical digital signal. Choosing the ratio between the two optical intensities is a critical design parameter. This parameter, which will be referred to as the extinction ratio, affects system performance by changing the receiver sensitivity. Figure 9 shows the loss in receiver sensitivity due to finite extinction ratios. An infinite extinction ratio occurs when no bias current is supplied that causes the ILD power to be turned off during a space. Unfortunately, this situation also results in a turn-on delay of several nanoseconds. When the bias current is below threshold,

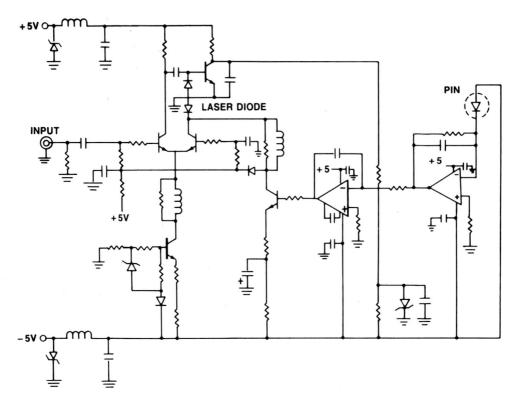

FIGURE 8.   ILD modulator circuit.

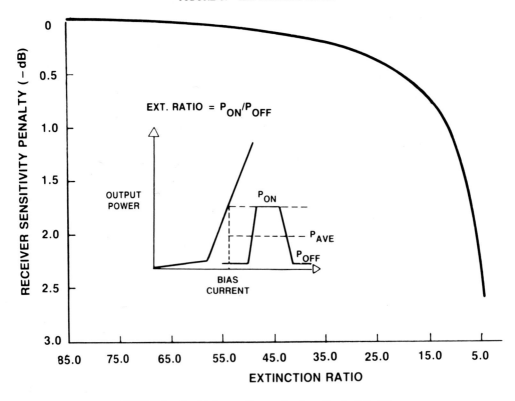

FIGURE 9.   Sensitivity penalty as a function of extinction ratio.

some time is needed for the carrier concentration to build up to the level associated with the threshold current. This turn-on delay causes a pulse pattern dependency in the light output and limits the maximum modulation rate of the laser. The highest modulation rates can be achieved by biasing the laser above threshold, but this degrades the extinction ratio. A good compromise is biasing the laser at about 95% of the threshold current; this produces an extinction ratio of better than 10:1 and allows modulation rates of several hundreds of Mb/sec.

It should be mentioned that laser diodes show a significant increase in noise (above the normal quantum noise level) at drive currents close to either side of threshold. Thus biasing too close to threshold should be avoided. The selection of the bias current also influences the spectral output and transient behavior of the laser. Figure 10 shows the optical pulse generated by a fixed modulation current and two different bias currents. Also shown are the detected frequency spectra generated by the laser. The damped relaxation oscillations at the leading edge of the pulses in Figure 10 are caused by interactions between carriers and photons. When a fast rise time pulse is applied to the laser, the carrier density overshoots and generates a strong optical field that depletes the carrier concentration below the level associated with the drive current and another overshoot follows. Usually, the ringing damps out rapidly. The frequency, magnitude, and damping coefficient of the oscillations are related to the bias current. The oscillations occur at frequencies in the 1 GHz region.

Earlier in this section we stated that there are three significant noise phenomena associated with interactions between laser and fiber. Optical feedback noise[32] is the result of varying optical feedback into the laser that changes the laser characteristics and has been discussed in Chapter 4, Section III.D. Mode partition noise,[22-25] which is caused by amplitude fluctuations of the longitudinal lasing modes and chromatic dispersion in the fiber, was described earlier in this section. Modal noise[30,31] is caused by the combination of mode dependent optical losses and changes in phase between modes or fluctuations in the distribution of energy among the modes. It is self-evident that the combination of mode dependent loss and fluctuations in the energy of individual modes will cause undesired changes (noise) in the power output. It is less obvious why the combination of differential mode loss and changes in phase between modes would have the same effect. If light from a coherent laser is launched into a multimode fiber, a speckle pattern results from the interference between propagating modes. Any change in phase between the modes will change the interference and will result in a different speckle pattern. Differential mode attenuation will cause losses to be dependent on the speckle pattern where the differential mode attenuation occurs. Typical sources of differential mode attenuation, and thus system losses that depend on speckle pattern, are splices, connectors, and microbending. Another less obvious source is a detector with a non-uniform responsivity or an overilluminated detector. Changes in phase between modes are caused by vibrations of the fiber (microphonic effect) and by frequency changes of the source.

As the light propagates through the fiber, a reduction in speckle contrast will take place. As the speckle pattern becomes less granular, less modal noise is generated by misaligned splices and other sources of differential mode loss. This indicates that modal noise can be reduced by using low coherence multimode lasers.[30] The use of a truly single mode waveguide will totally eliminate modal noise.* Experience has shown that modal noise is a minor problem in digital systems that were designed with a reasonable power margin.

This section has discussed typical solid state optical sources used in current fiber optic transmission systems. The sources available today have been used in optical transmission systems that demonstrate definite improvements economically and technically over wire and coaxial systems. However, from a communication theory point of view, we are using noise carrier transmission systems.

---

* A conventional single-mode fiber actually supports two orthogonally polarized modes.

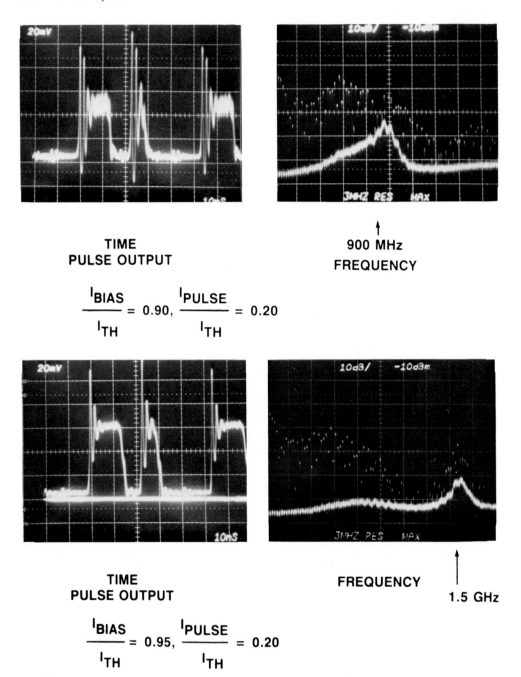

FIGURE 10. Time and frequency domain representation of a detected ILD optical pulse.

The spectral purity and carrier frequency stability of present optical sources are inadequate for coherent ASK, PSK, and FSK transmission. Research is underway to develop laser structures or techniques to improve the carrier stability and purity in optical sources so that the full potential of optical fiber bandwidth can be achieved.

IV. OPTICAL FIBER WAVEGUIDE CONSIDERATIONS

In this section, we will provide a brief overview of the properties and important parameters

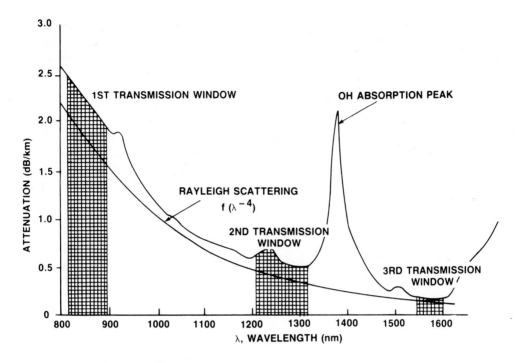

FIGURE 11.   Optical fiber attenuation as a function of wavelength.

that characterize optical fibers. Readers who prefer a thorough mathematical treatment of optical waveguides are referred to Chapters 1 and 3 and the references.[35-37]

When considering the application of fiber optic waveguides for a digital communication link, the acceptable insertion loss and bandwidth necessary to traverse the distance must be determined. There is a choice of fiber types that satisfies a wide range of digital communication needs. The fiber index of refraction profile and the operating wavelength are two basic parameters to be considered. Each profile class and wavelength regime has a distinct characteristic relevant to the information rate and transmission distance of the communication system.

Figure 11 shows the dependence of fiber attenuation on wavelength. There are three principal operational wavelength windows useful for communication purposes. The first window falls in the 810 to 850 nm region. Early fiber optic systems utilized this region almost exclusively since optical sources, detectors, and fiber technology were all wavelength compatible. The emission wavelength of AlGaAs sources fell within the most responsive region of silicon photodetectors. Fiber manufacturers were able to produce fibers with 6 dB/km and lower attenuation in the first wavelength window. In early 1980, optical sources and detectors, using InGaAs and InGaAsP materials, which operate in the 1200 to 1600 nm region, became available. This long wavelength region allows operation with lower loss and lower dispersion in fibers.

Attenuation in optical fiber waveguides is the result of four basic mechanisms: scattering, absorption, microbending, and macrobending. Scattering is introduced by an intrinsic mechanism identified as Rayleigh scattering. Rayleigh scattering losses are inversely proportional to the fourth power of wavelength. Other scattering type losses may be introduced by random inhomogenities in the glass. Rayleigh scattering is the dominant loss mechanism in the best available fibers.

Absorption losses are introduced by OH ions and metallic impurities introduced into the glass during the manufacturing process. Infrared and ultraviolet absorption of the glass material itself will add to the loss. The peaks in the attenuation characteristics in Figure 11 are absorption loss mechanisms, and the gradual (λ^{-4}) decrease is due to Rayleigh scattering.

The process of protecting the fiber[33] by coatings and cable structures will introduce what is referred to as microbending. Microbends are microscopic perturbations along the fiber's longitudinal axis. Microbending causes mode mixing and can cause significant radiative losses. Macrobends cause higher order modes to convert to nonguided or radiation modes when encountering a visible-to-the-eye bend. Macrobends usually do not cause significant losses as long as one adheres to the manufacturer's specification for minimum bending ratios. Other losses that must be accounted for in a fiber link include splicing and connector losses. Splices are made by mechanically attaching two fiber ends and applying an index matching epoxy. The spliced fibers are secured in a protective enclosure. Fibers are also spliced by fusing the fiber ends by melting the glass with an electric arc. Mechanical and arc fusion splices generally have losses between 0.1 and 0.5 dB. A lower limit on loss can be attributed to the eccentricity, ellipticity, core diameter, and NA profiles. The (multimode) step-index fiber is limited to low bit rate, short distance communication systems. The most significant limitation is the relatively large differential mode delay. Differential mode delay can be rather easily explained using geometrical optics (rays). In geometrical optics, modes are visualized as rays passing through the longitudinal axis of the fiber.* An individual mode is identified by the angle its ray makes with the fiber axis. Higher order modes pass through the axis at larger angles. At the interface between core and cladding a ray will be totally reflected, maintaining the same angle with the axis, as long as the angle at which it intersects the fiber axis is sufficiently shallow. The maximum angle at which total reflection takes place is obtained using Snell's law of refraction:

$$n_1 \cos\theta_{max} = n_2 \tag{8}$$

where n_1 is the index of refraction of the core, n_2 is the (lower) refractive index of the cladding and θ_{max} is the angle between the ray and the fiber axis at which the reflected ray emerges parallel to the core cladding interface.

Since rays with different angles have different path lengths, their propagation time will be different. The propagation time is simply given by

$$t_p = \frac{n_1 L}{c \cos\theta_r} \tag{9}$$

where t_p is the propagation time of mode p, seconds; n_1 is the core region index of refraction; c is the velocity of light in a vacuum, meters/seconds; θ_r is the angle the ray makes with the fiber's longitudinal axis; and L is the fiber length, meters.

The differential mode delay, Δt, then is the difference between the propagation time of the lowest order mode ($\theta_r = 0$ radians) and the highest order mode ($\theta_r = \theta_{max}$)

$$\Delta t = t_{max} - t_o = \frac{n_1 L}{C} \left(\frac{n_1}{n_2} - 1 \right) \tag{10}$$

This measure of pulse broadening differs from the usual rms pulse broadening, σ which is derived from the characteristics of the impulse response, h(t), of the fiber.

---

* Such rays are called meridional rays. A simplification has been made by ignoring skew rays which encircle the fiber axis. Ignoring skew rays does not substantially affect the conclusions made about fiber characteristics. Skew rays are discussed in more detail in several references.[34,35]

$$\sigma^2 = \int_{-\infty}^{\infty} h(t)\,(t - \bar{t})dt$$

$$\text{where} \quad \bar{t} = \int_{-\infty}^{\infty} h(t)t\,dt \text{ and } \int_{-\infty}^{\infty} h(t)dt = 1 \tag{11}$$

We will relate the pulse broadening to two commonly used fiber parameters, namely the refractive index contrast, Δ, which is given by

$$\Delta = \frac{n_1 - n_2}{2n_1^2} \cong n_1 - \frac{n_2}{n_1} \tag{12}$$

and the numerical aperture (NA) which is given by

$$NA = (n_1^2 - n_2^2)^{1/2} = n_1(2\Delta)^{1/2} \tag{13}$$

Using Equations 10, 12, and 13, the pulse broadening due to modal dispersion is simply given by

$$\Delta t = \frac{n_1}{c}\left(\frac{n_1 - n_2}{n_2}\right) = \frac{n_1}{c}\,\Delta \text{ nsec/km} \tag{14}$$

where c is the speed of light in km/nsec ($c = 3 \times 10^{-4}$)

It is evident from the above equations that reduction of Δ will reduce both the pulse broadening and the NA. A large NA facilitates source-fiber coupling (see discussions in Chapter 3, Section V and Chapter 4, Section III.C.). Thus the selection of Δ will be a compromise between the pulse broadening in the fiber and the source to fiber coupling efficiency.

Modal dispersion can vary between 10 nsec/km and as high as 50 nsec/km for step-index fiber. Mode mixing in the fiber and at splices or connectors makes an exact dispersion value difficult to predict.

An effective mechanism to reduce modal dispersion is grading of the refractive index of the core so that the index decreases in a near parabolic fashion radially away from the fiber axis. A fiber with this type of profile is called a graded-index fiber. In a graded-index fiber, higher order modes propagate faster than the lower order modes; therefore, the longer path length of the higher order modes is offset by the higher average group velocity. There is no index profile that perfectly equalizes the group velocity of all modes in a multimode circular symmetric fiber. The commonly used near parabolic profile is referred to as the power-law index profile or the alpha profile and is characterized by the following equation.

$$n(r) = n_1[1 - 2\Delta(r/a)^\alpha]^{1/2} \quad r \leq a \tag{15}$$

$$n(r) = n_2 = n_1[1 - 2\Delta]^{1/2} \quad r \geq a \tag{16}$$

where a is the radius of the core and n(r) is the refractive index as a function of radius.

Multimode distortion is minimized by selecting a particular value of α which depends on fiber material and operating wavelength. For most materials, this optimum value is around two. The derivation of the optimum value of α can be found in the References 34 and 36. The modal dispersion of practical multimode fibers is determined by unavoidable deviations from the optimum profile. Typical values for modal dispersion of graded-index fibers manufactured in 1982, range from 20 to 300 psec/km.[38] The number of guided modes in an α profile graded-index fiber is approximately half of that in a step-index fiber with the same NA.

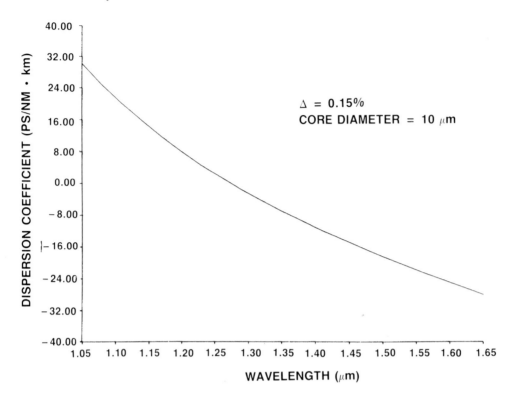

FIGURE 12.   Material dispersion vs. wavelength.

A. Chromatic Dispersion

Chromatic dispersion is the term used to describe the distortion caused by the dependence of the propagation characteristics of a mode on wavelength. Three different effects contribute to chromatic dispersion, namely, material dispersion, waveguide dispersion, and profile dispersion. Of these three components, material dispersion normally dominates at wavelengths between 800 and 900 nm. Material dispersion refers to that part of the chromatic dispersion that is attributable to the wavelength dependence of the refractive index. Material dispersion is characterized by the material dispersion parameter D as discussed in Chapter 3, Section III. Figure 12 shows the dependence of the material dispersion parameter on wavelength. Pulse broadening due to material dispersion in a unit length of optical fiber is obtained by multiplying the material dispersion coefficient D (nsec/nm/km) with the spectral linewidth $\Delta\lambda$, of the source:

$$\sigma_{material} = \frac{\lambda\Delta\lambda}{c} \frac{d^2n}{d\lambda^2} \tag{17}$$

where λ is the source center wavelength, $\Delta\lambda$ is the source spectral width (RMS), n is index of refraction of the core, and c is the velocity of light.

At 820 nm a typical material dispersion coefficient is about 100 psec/nm/km. The second derivative of the refractive index with respect to wavelength passes through zero in the 1200 to 1500 nm region.[*]

Waveguide dispersion is caused by the fact that the propagation characteristics of a mode are dependent on the geometric properties of the waveguide as a function of wavelength.

---

[*]   As mentioned in Chapter 3, Section III, at this zero dispersion wavelength, pulse broadening due to the residual dispersion D_2 (nsec/km/nm^2) becomes important.

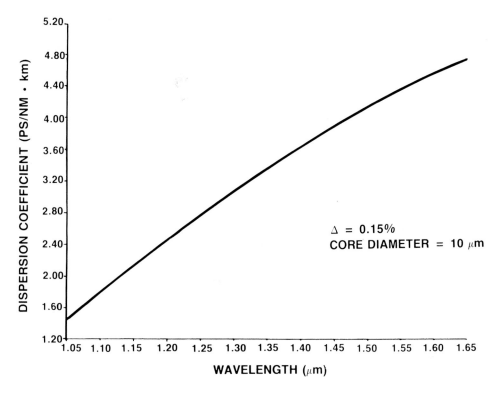

FIGURE 13. Waveguide dispersion vs. wavelength.

For circular waveguides, the dependence is on the ratio between core radius and wavelength. Waveguide dispersion is normally only of interest in single mode wave guides in the vicinity of the zero (material) dispersion wavelength. Figure 13 shows the dependence of the waveguide dispersion on wavelength in a single mode waveguide.

Profile dispersion is caused by variation of the refractive index profile with wavelength, which is caused by variation of the refractive index contrast with wavelength and/or variation of the profile parameter with wavelength. In multimode waveguides, the optimum profile (i.e., the profile that minimizes differential mode delay) is dependent on the profile dispersion parameter P^{30} which is defined by:

$$P(\lambda) = \frac{n_1}{N_1} \frac{\lambda}{\Delta} \frac{d\Delta}{d\lambda} \tag{18}$$

where n_1 is the refraction index at the core axis, $N_1 = n_1 - \lambda(dn_1/d\lambda)$ is the group index, Δ is the refractive index contrast, and λ is the operating wavelength.

In single mode waveguides, material, waveguide, and profile dispersion are interrelated in a complicated way and depend on the dispersive effects of both core and cladding.[40] Figure 14 shows the dependence of the total dispersion and its components on wavelength in a single-mode waveguide. Analysis has shown that the shape of the total dispersion curve and the zero dispersion wavelength depend strongly on the refractive index profile geometry.[41,42]

B. Bandwidth of Concatenated Fibers

Fiber cable manufacturers rarely, if ever, give the values of dispersion for multimode fibers. A more typical specification given is the bandwidth-length product. This specification is the optical half power modulation bandwidth for a continuous, 1 km length of fiber.

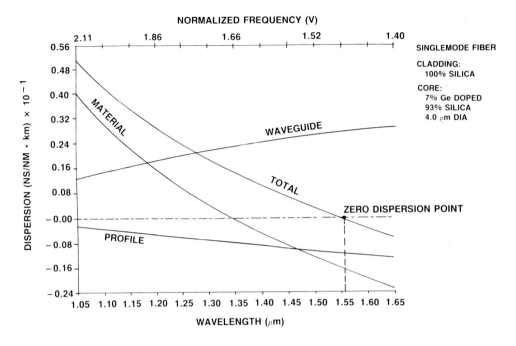

FIGURE 14.   Dispersion vs. wavelength for a single mode fiber.

Calculation of the end-to-end fiber bandwidth from the bandwidth of the individual fiber sections is unfortunately not very precise due to mode mixing at splice points. Two limits can be set by assuming complete mode mixing and a second case where no mode mixing occurs. In the former case, the bandwidth reduces inversely proportional to the square root of the number of fiber sections. The latter case results in bandwidth reducing inversely to the number of fiber sections. The total bandwidth of sections of fiber should be governed by the following expression.

$$\frac{BW_o}{\ell} \leq BW_{total} \leq \frac{BW_o}{\ell^{1/2}} \tag{19}$$

where BW_o is the bandwidth of 1 km length of fiber, and ℓ is the number of 1 km lengths of fiber.

The inequality implies that the total bandwidth could be inversely proportional to length raised to an exponent between 0.5 and 1.0. That is:

$$BW_{total} \approx \frac{BW_o}{\ell\gamma} \tag{20}$$

where γ, is the bandwidth reduction exponent.

When mode mixing is minimal, errors in the profile of graded-index fibers can offset each other, which also would lead to a less than linear decrease in bandwidth.

Curves of bandwidth reduction for exponents of 0.5 and 1.0 are shown in Figure 15. Also shown are a set of points at 1 km intervals showing bandwidth measurements made on an actual field installed cable. It can be seen that the bandwidth behavior can be quite erratic, in some cases actually increasing as an extra fiber is spliced. Most manufacturers will supply the expected value of γ for their fibers. It should be stressed, however, that this is not a reliable system parameter,[43] and that there is some risk that the actual end-to-end bandwidth might be less than the calculated one.

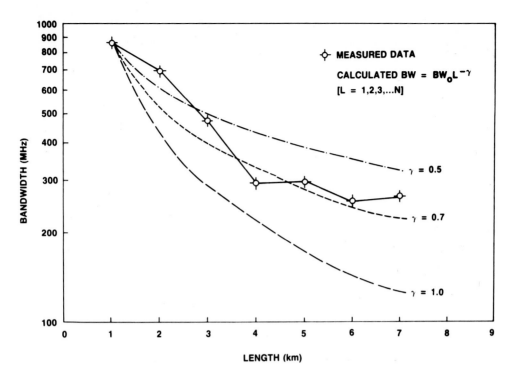

FIGURE 15. Bandwidth of concatenated fiber sections.

V. RECEIVER CONSIDERATIONS

The function of the receiver in an optical fiber link is to convert the optical signal back to an electrical signal. Performance of the receiver is usually judged by its sensitivity to the incident optical power. The receiver sensitivity is defined as the minimum amount of optical power required to obtain a specified BER. In the following discussion, we will restrict ourselves to communication systems where binary digits are transmitted. Furthermore, we will assume that the probability that a "1" (mark) was sent equals the probability that a "0" (space) was sent.

Figure 16 shows a block diagram of a typical receiver. An optical detector converts the optical signal to an electrical current. Typically this curent is weak and is amplified by a low noise preamplifier followed by a low pass filter, gain stages, and a decision circuit.

Typically, only two types of devices are employed for the detection of an optical signal, either simple PIN diodes or more sophisticated avalanche photodiodes (APD) with internal current gain. APDs are most compatible with optical fiber systems and have shown best performance thus far.

With PIN diodes, the fundamental noise limitation is usually referred to as quantum noise.*[44] Quantum noise is caused by the statistical nature of the detection process. In a reverse-biased photodiode incoming photons are absorbed, and hole-electron pairs are generated within the device. These photogenerated carrier pairs separate under the influence of the internal detector fields and produce a displacement current. On the average, the number of carrier pairs generated per second, $<c(t)>$, is related to the incident optical power, $p(t)$, by the quantum efficiency, η, of the detector.

---

* The background radiation, which is considered the fundamental limit for microwave receivers, is very small and can be neglected.

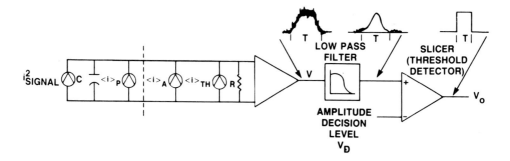

FIGURE 16. Simplified receiver diagram.

$$<c(t)> \ = \ \eta \ p(t)/h\nu \tag{21}$$

where $h\nu$ is the energy in a photon.

Thus, for a given incident optical power, $p(t)$, we can calculate the average rate at which carrier pairs are generated. However, we cannot predict the actual number of electron-hole pairs that will be generated in a time interval T. The probability that exactly C carrier pairs will be generated is given by the Poisson distribution.[45]

$$P(c = C) \ = \ e^{-\mu} \frac{\mu^c}{C!} \tag{22}$$

where $\mu \ = \ \eta/h\nu \int_T p(t)dt$ and $\int_T p(t)dt$ is the energy received in time interval T.

Equation 22 shows that there is a finite probability that zero carrier pairs will be generated when an optical pulse is present. Thus the sensitivity of a receiver is limited by the statistics of photodetection, and, assuming a receiver able to observe a single carrier pair, one can calculate the minimum required pulse energy, referred to as the quantum limit, to achieve a particular BER. It can easily be shown that the minimum detectable power level for a BER of 10^{-9} is given by

$$p_{min} \ = \ 10.5 \ h\nu B/\eta \qquad \text{Watts} \tag{23}$$

where B is the bit rate.

In an actual PIN diode, shot noise from surface leakage and dark currents is added to the quantum noise.

The avalanche process in an APD not only results in a gain of the signal current and associated quantum noise, but because of the statistical nature of the multiplication process, there is a gain distribution that gives rise to excess noise. The excess noise increases with increasing gain and thus eventually limits the magnitude of the useful gain. As long as the noise generated by the avalanche photodiode is small in comparison to the noise added by the preamplifier, a net improvement in receiver sensitivity is observed with increasing gain. The avalanche gain distribution and excess noise term are dependent upon the ionization ratio k, which describes the ionization rates of holes vs. electrons.

In addition to the excess noise, the useful gain in an avalanche photodiode is also limited by dark current, which may undergo multiplication.

The amplification that follows the photo detection adds additional noise that often dominates the quantum noise generated in the detector. Therefore, the sensitivity of a receiver, especially using a PIN photodiode, will be in a large part determined by the noise characteristics of the preamplifier. The noise sources in the preamplifier are generally assumed to be independent and to have Gaussian statistics.

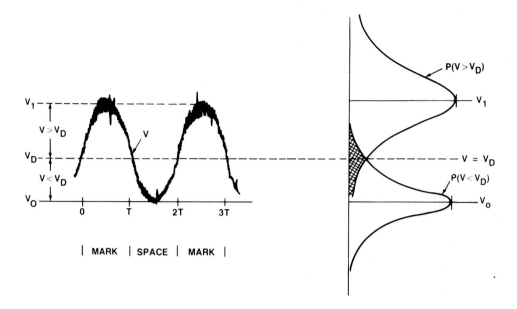

FIGURE 17.   Probability distribution function for a noisy, two-level signal.

The noise added to the signal causes the two possible states of the signal to be represented by probability density functions. The mean values of the two states are "0" and "1" as depicted in Figure 17. The shapes of the distribution are not identical, which is the result of signal level dependent detector quantum noise. A solution is generally found that establishes equal incidence of errors at both states. With equal noise at the two levels of the digital signal, it would be expected that the decision level, or threshold, would be at the 50% point. However, since the noise is signal dependent, the optimum threshold will occur at somewhat less than 50%. Since the area under the probability density functions is unity, meaning an event did occur, an assessment of the error rate can be accomplished by finding that portion of the area under the curve that represents the tails of the respective probability density function beyond the threshold level (the cross-hatched portion in Figure 17). Since the probability density function represents several noise sources with different characteristics, difficult and tedious calculations are required in order to evaluate error rate performance exactly, or to determine bounds to error rate. The results of such calculations have shown little deviation from calculations that assumed all of the noise sources to have Gaussian statistics.

A simple model for an optical receiver is shown in the illustration of Figure 16 (noise model). The model consists of a detector, amplifier, a filter network, and a decision circuit. Current sources for the modeled detector consist of the detected signal, signal-dependent quantum noise, and dark current. The equivalent impedance of the detector is the depletion capacitance plus parasitic packaging capacitance.

Noise current generators for the amplifier are needed for thermal noise associated with physical resistances and shot noise term associated with the input device, base or gate, and, collector or drain, terminals. The active device noise is usually frequency dependent so that device selection becomes a design consideration for different bit rates.

The detector preamplifier is followed by a low pass filter in designs not integrating the received signal. Optimization of the signal to noise ratio may be achieved by a correct selection of band limiting filter. If for instance an NRZ signal is received, 90% of the total energy is contained within a bandwidth numerically equal to the bit rate. An optimum SNR can ideally be achieved with a Nyquist filter with a cut-off frequency equal to one half the bit rate.[46] The design of such an ideal Nyquist filter would require an infinite number of filter sections, which is obviously impractical. Practical filters can be implemented anywhere from 20 to 30% wider in bandwidth than an ideal Nyquist filter. The amplitude function of a Nyquist filter for NRZ pulses is given by

$$
H(f) = \begin{cases} \dfrac{\pi f/B}{\sin(\pi f/B)} & \left[0 \le f \le \dfrac{B}{2}(1-\alpha) \right] \\[3mm] \dfrac{\pi f/B}{\sin(\pi f/B)}\cos^2\left\{\dfrac{\pi}{4\alpha B}\,[2f - B(1-\alpha)]\right\}, & \left[\dfrac{B}{2}(1-\alpha) \le f \le \dfrac{B}{2}(1+\alpha) \right] \\[3mm] 0 & \left[f > \dfrac{B}{2}(1+\alpha) \right] \end{cases}
\tag{24}
$$

with $0 \le \alpha \le 1$ and $\alpha = 0$ denoting the ideal Nyquist filter.

Bandlimiting the NRZ signal produces a waveform approximating a raised cosine. Figure 17 (raised cosine shape and noise) shows a bandlimited NRZ sequence corrupted by noise. The effect on noise may be visualized statistically as expected values at level V_o and level V_1. Noise added to the signal results in an uncertainty in the ability to distinguish the received level.[47] The uncertainty is represented in Figure 17 by a normal or Gaussian distribution about each level. An optimum decision level V_D is established at the intersection of the two distributions. Moving the decision level to either side of the optimum will result in a greater number of wrong decisions or a degraded bit error rate. The probability of error can be stated as follows

$$
P(v > v_D) = \frac{1}{2\pi\sigma_0^2} \int_{v_D}^{\infty} \exp\left[-\frac{1}{2}\frac{(v)^2}{\sigma_0^2} \right] dv
\tag{25}
$$

$$
P(v < v_D) = \frac{1}{2\pi\sigma_1^2} \int_{-\infty}^{v_D} \exp\left[-\frac{1}{2}\frac{(v - v_1)^2}{\sigma_1^2} \right] dv
\tag{26}
$$

$$
P(v > v_D) = P(v < v_D)
\tag{27}
$$

where v_1 is the signal peak amplitude and σ^2 is the variance of the added Gaussian noise. Another interpretation is that σ^2 is the noise power level.

The value of σ^2 in an optical receiver can be calculated by considering the model shown in Figure 16. A list of the noise powers is shown below.

1.   Signal-dependent quantum noise[48]

$$
<i_\varphi^2> = 2q\left(\frac{nq}{h\nu}\right)\overline{P}_R <g>^2 F_x B_N
\tag{28}
$$

2.   Dark current noise (bulk)

$$
<i_D^2> = 2qI_D <g>^2 F_x B_N
\tag{29}
$$

3.   Thermal noise[49]

$$
<i_T^2> = \frac{4kTB_N}{R}
\tag{30}
$$

4.   Expressions for the active device noise current source behavior are available in the literature and are often given by commercial manufacturers.

Where q is electron charge, 1.6×10^{-19} C; $\overline{P}_R$ is average received power, watts; g^2 is mean square gain ($<g^2> = 1$ for a PIN diode); k is Boltzman's constant, 1.38×10^{-23} J/K; T is temperature in degrees kelvin; and I_D is detector dark current, amperes.

The detected signal power of a random NRZ waveform is

$$\frac{1}{2}\left[2\left(\frac{x-1}{x+1}\right) \frac{\eta q}{h\nu} P_R <g> \right]^2 \tag{31}$$

The expressions for signal and noise power involve a mean gain term, $<g>$, which equals unity for PIN-type detectors. In the case of APD detectors, the mean gain takes on values on the order of tens to hundreds. Noise power is also effected by the avalanche gain mechanism, which is expressed by an excess noise factor F_x. The excess noise factor also becomes unity for PIN detectors. F_x can be expressed as:

$$F_x = k <g> + (2 - <g>^{-1})(1 - k) \tag{32}$$

where k is the ratio of electron and hole ionization rate; k is about 0.035 for silicon APDs and 1.0 for germanium APDs.

The receiver sensitivity may be expressed using a combination of the above expressions.

$$P_R = \left\{ \left[\frac{q <g^2> B_N}{2\left(\frac{x-1}{x+1}\right)^2} \frac{h\nu}{\eta q} \cdot snr \right] \right.$$

$$+ \frac{1}{2}\left\{ \left[\frac{q <g^2> B_N}{4\left(\frac{x-1}{x+1}\right)^2 <g>^2} \left(\frac{h\nu}{\eta q}\right) \cdot snr \right]^2 \right.$$

$$\left. + \frac{2S_A B_N}{\left(\frac{x-1}{x+1}\right)} \frac{h\nu}{\eta q} \cdot snr \right\}^{1/2} \tag{33}$$

where P_R = received average power, x = source extinction ratio, h = Plank's constant, ν = optical frequency, q = charge of an electron, η = detector quantum efficiency, snr = signal-to-noise ratio for the desired BER, $<g>$ = detector mean gain, $<g^2>$ = detector mean square gain (note $<g^2> = <g>^2 F_x$), F_x = detector excess noise factor, S_A = preamplifier input current noise spectral density, and B_N = noise bandwidth.

The received power in many cases can be minimized if an APD detector is used where the preamplifier noise dominates in the determination of snr. The mean gain, $<g>$, of an APD is adjustable so that the signal current may be increased. The optimum adjustment of gain occurs when the APD excess noise equals the preamplifier noise. Any further gain increase results in a deteriorating snr. Figure 18 shows graphically the effects of detector gain adjustment on the required power to achieve a stated BER (or snr). The minimum shown on the curve defines the receiver sensitivity factor. Of course, for a PIN diode detector,

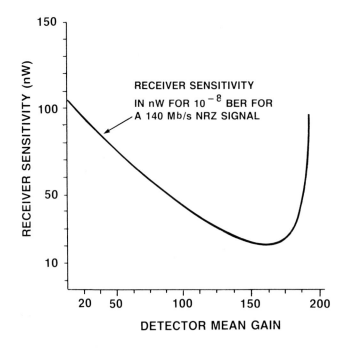

FIGURE 18. Receiver sensitivity vs. avalanche gain.

gain optimization is not available. When designing PIN receivers, the only option available for improving the receiver sensitivity is to select a preamplifier design with better noise performance.

As described above, the sensitivity of a receiver will to a large extent be determined by the choice of a preamplifier. In many applications, receiver sensitivity is not the only consideration for the selection of a preamplifier, and, often, dynamic range is an equally important criterion. It is common to distinguish preamplifier designs into three generic types: namely, the resistively loaded input amplifier, the integrating or high impedance amplifier, and the transimpedance amplifier.

The resistively loaded input amplifier consists of a load resistance that bears the detector signal current. The voltage generated across the resistor is amplified in most cases by a 50Ω impedance amplifier.

In many cases, the load resistor is used to match the 50Ω amplifier as well as to develop the detected signal voltage. The prime advantage of this approch is the ease of providing wide bandwidths; the disadvantage is the relatively poor noise performance. The noise spectral density is determined by the thermal noise of the 50Ω resistor and the noise figure of the 50Ω amplifier.

$$<i_A^2> = \frac{4k\ T_{eff}}{R_{50}} \tag{34}$$

where

$$T_{eff} = 290\left(10^{\frac{NF}{10}} - 1\right)\ ,\quad °K \tag{35}$$

and NF amplifier noise figure in dB.

Although this type of amplifier has a relatively poor noise performance, it might be the only design alternative that provides the bandwidth for bit rates extending into the Gb/sec region.

The integrating front end is a photodetector supplying signal current into a high input impedance amplifier. Having a high physical resistance results in a low level input current spectral noise density, and consequently this amplifier produces the most sensitive receivers. However, there are several drawbacks to the integrating front-end design that have limited its applications. Foremost is the intersymbol interference created by the high input RC time constant that must be compensated for by use of post amplifier equalizer networks. (In some cases the RC time constant is determined by parasitic capacitance that requires a unique equalizer network.) Second, long periods without signal transitions result in baseline wander (DC drift). This action may cause signal saturation internal to the preamplifier, and at a minimum reduces the receiver's dynamic range. Last, the nature of the amplifier high impedance allows the designs not only to be sensitive to detected signal currents, but also to undesired RFI and EMI.

Currently, the most popular design is the transimpedance amplifier. As its name implies, the signal current entering the input post of the amplifier is converted to a voltage at the output port. This design is implemented by closing a resistive feedback loop from the output back to the input. The value of the feedback resistance is limited by the requirement for adequate loop gain around the feedback path. The first active stage is usually in either a grounded emitter configuration (bipolar transistor) or as a grounded source (field effect transistor). The physical input resistance of the first stage is quite high, while the effective resistance is reduced by the amplifier's open loop gain. The low effective or apparent resistance of the input reduces the amount of intersymbol interference (eliminating the need for equalizers). The transimpedance ratio of the amplifier is set by the value of the feedback resistance.

Noise sources within the transimpedance amplifier include the feedback resistor thermal noise and the shot noise generated by the base current (gate current) and the collector current (drain current) of the input transistor (FET).

The transimpedance amplifier provides a good compromise between receiver sensitivity and dynamic range. A comprehensive review of receiver theory is available in the literature.[50]

REFERENCES

1. **Basch, E. and Hanna, D.,** GTE Long Wavelength Optical Fiber Communications System, *I.C.C. Rec.,* ICC'82 5D4.1-5D4.5, Philadelphia, June 13 to 17, 1982.
2. **Basch, E. and Carnes, H.,** Investigations in long wavelength fiber systems, *National Engineering Consortium,* 36, 17, 1982.
3. **Cancellieri, G. and Ravaioli, U.,** Base band response of multimode optical fibers under different excitation conditions, *J. Opt. Commun.,* 3 (1), 13, 1982.
4. **Grothe, H., et al.,** Transmission experiments and system calculations using 1.3 μm LED, *J. Opt. Commun.,* 3 (2), 63, 1982.
5. **Garrett, I. and Todd, C. J.,** Components and systems for long-wavelength monomode fibre transmission, *Opt. Quant. Electron.,* 14, 95, 1982.
6. **Yamamoto, Y. and Kumura, T.,** Coherent optical fiber transmission systems, *IEEE J. Quant. Electron.,* QE17 (6), 919, 1981.
7. **Shapiro, J. H.,** Earth Space Optical Communication through Atmospheric Turbuence: Theory, *I.C.C. Rec.,* ICC'79 1.11—1.13, Boston, June 10 to 14, 1979.
8. **Basch, E., Carnes, H., and Kearns,** Heterodyning system research: update, *OFC Tech. Dig.,* 56, 58, 1983.
9. **Hodgkinson, T. G., Wyatt, R., Malyon, D. J., Nayar, B. K., Harmon, R., and Smith, D. W.,** Experimental 1.5 μm coherent optical fiber transmission system, 8th European Conference on Optical Communication, Cannes, September 21 to 24, 1982, 414.
10. **Kikuchi, K. and Kemura, T.,** Achievement of nearly shot-noise-limited operation in a heterodyne-type in PCM-ASK optical communication system, 8th European Conference on Optical Communication, Cannes, September 21 to 24, 1982, 419.

11. **Gardner, F.,** *Phase Lock Techniques,* John Wiley & Sons, New York, 1979.

12. Members of Technical Staff, Bell Telephone Laboratories, Transmission Systems for Communications, Western Electric Company, Inc., Dec. 1971, 659.

13. **Duttweiler, D. L.,** The jitter performance of phase locked loops entracting timing from baseband data waveforms, *BSTJ,* 55, 37, 1976.

14. **Feher, K.,** Digital Modulation Techniques on an Interference Environment, Don While Consultants, Inc., Germantown, Md., 1977, p. 3.1.

15. Technical Staff of CSELT, Optical Fiber Communications Centro Studie Laboratori Telecommunications, Turin, 1980, 703.

16. **Kasper, B. L.,** Equalization of multimode optical fiber systems, *BSTJ,* 61 (7), 1367, 1982.

17. **Personick, S. D.,** Receiver design for digital fiber optic communication systems I, *BSTJ,* 52, 843, 1973.

18. **Smith, D. R. and Garret, I.,** A simplified approach to digital optical receiver design, *Opt. Quant. Electron.,* 10, 211, 1978.

19. **Millman, J. and Taub, H.,** *Pulse, Digital and switching Waveforms,* McGraw Hill, New York, 1965.

20. **Miller, S. E. and Chynoweth, A. T., Eds.,** *Optical Fiber Telecommunications,* Academic Press, New York, 1979, 557.

21. **Sueta, T. and Izutsu, M.,** High speed guided-wave optical modulators, *J. Opt. Commun.,* 3 (2), 52, 1982.

22. **Ogawa, K.,** Analysis of mode partition noise in-laser transmission systems, *IEEE J. Quant. Electron.,* May 1982, p. 849.

23. **Okana, Y., Nagama, K., and Ito, T.,** Laser mode partition noise evaluation for optical fiber transmission, *IEEE Trans. Commun.,* COM-28, 238, 1980.

24. **Elze, G. and Patzak, E.,** Modal mode partition noise in optical fiber systems, *J. Opt. Commun.,* 3 (2), 67, 1982.

25. **Kopf, G. G., Kuller, L., and Patzak, E.,** Laser mode partition noise in optical wideband transmission links, *Electron. Lett.,* 18, 493, 1982.

26. **Lagasse, P. E. and Timmerman, R.,** Optical fiber communication systems using LED's as light sources, *Electro-Optics Laser, '78; Proc. Tech. Program,* September 19 to 21, 1978, Boston.

27. **Basch, E. E., and Carnes, H. A.,** Aspects of an Operational Fiber Optic System, *I.C.C. Rec.,* ICC'79 pp. 19.6.1-19.6.7, 1979.

28. **Lee, I. T. P., Burrus, C. A., Copeland, J. A., Dentai, A. G., and Marcuse, D.,** Short-cavity InGaAsP injection lasers: dependence of diodes spectra and single-longitudinal mode power on cavity length, *IEEE J. Quant. Electron.,* QE 18, 1101, 1982.

29. **Goldberg, L., Dandridge, A., Miles, R. O., Giallorenzi, T. G., and Weller, J. F.,** Noise characteristics in line narrowed semiconductors lasers with optical feedback, *Electron. Lett.,* 17 (19), 677, 1981.

30. **Epworth, R. E.,** Modal noise-causes and cures, *Laser Focus,* 17(9), 109, 1981.

31. **Rawson, E., Goodman, J., and Norton, R.,** Experimental and analytic study of modal noise in optical fibers, 6th European Conference on Optical Communication, York, U.K., September 1980, 72.

32. **Petermann, K.,** Nonlinear distortions and noise in optical communication systems due to fiber connectors, *IEEE J. Quant. Electron.,* QE-16 (7), 761, 1980.

33. **Gloge, D.,** Optical-fiber packaging and the influence on fiber straightness and loss, *BSTJ,* 54 (2), 245, 1975.

34. **Okoshi, T.,** *Optical Fibers,* Academic Press, New York, 1982, 36.

35. **Barnoski, M. K., Ed.,** *Fundamentals of Optical Fiber Communication,* Academic Press, New York, 1976, 2.

36. **Midwinter, J. E.,** *Optical Fibers for Transmission,* John Wiley & Sons, 1976, 81.

37. **Okoshi, T.,** *Optical Fibers,* Academic Press, New York, 1982.

38. **Horigucki, M., et al.,** Transmission characteristics of ultra-wide bandwidth VAD fibers, 8th European Conference on Optical Communication, Cannes, September 21 to 24, 1982, 75.

39. **Sladen, F. M. E., Payne, D. N., and Adams, M. J.,** Profile dispersion measurements for optical fibers over the wavelength range 350 nm to 1900 nm, European Conference on Optical Communication, Genoa, September 12 to 15, 1978, 4.

40. **Gambling, W. A., Matsumura, H., and Ragdale, C. M.,** Mode dispersion, material dispersion and profile dispersion and graded-under single-mode fibers, *Microwaves Opt. Acoust.,* 3 (6), 239, 1979.

41. **Cohen, J. T., Mammel, W. L., and Lumish, S.,** Tailoring the shapes of dispersion spectra to control bandwidths in single-mode fibers, *Opt. Lett.,* 7 (4), 183, 1982.

42. **Lazay, P. D. and Pearson, A. D.,** Developments in single-mode fiber design, materials, and performance at Bell Laboratories, *IEEE J. of Quant. Electron.* QE-18 (4), 504, 1982.

43. **Reitz, P. R.,** Prediction of Length Performance of Multimode Graded-Index Fiber, Symp. Optical Fiber Measurements, Publ. 641, National Bureau of Standards, Boulder, Colo., Oct. 13, 1982.

44. **Oliver, B. M.,** Thermal and quantum noise, *Proc. IEEE,* 53, 436, 1965.

45. **Papoulis, A.,** *Probability, Random Variables and Stochastic Processes,* McGraw-Hill, New York, 1965, 101.

46. **Feher, K.,** *Digital Communications: Microwave Applications,* Prentice-Hall, New York, 1981, 47.
47. **Schwartz, M., Bennett, W., and Stein, S.,** *Communication Systems and Techniques,* McGraw-Hill, New York, 1966, 290.
48. **Webb, P., McIntyre, R., and Conradi, J.,** Properties of avalanche photodiodes, RCA Rev., 35, 244, 1974.
49. **Peebles, P.,** *Communication System Principles,* Addison-Wesley, Reading, Mass., 1976, 453.
50. **Smith, R. T. and Personick, S. P.,** Receiver design for optical fiber communication systems, *Top. Appl. Phys.,* 39, 89, 1982.

Chapter 7

ANALOG OPTICAL COMMUNICATIONS

E. E. Basch and H. A. Carnes

TABLE OF CONTENTS

I. INTRODUCTION

Information can be transmitted from one point to another in either digital or analog form. Despite the general trend toward digital transmission, there are many potential applications where analog fiber optic transmission systems compatible with other existing analog transmission systems are needed. Examples of analog use are found in the transmission of multiplexed video signals,[1-3] microwave radio,[4] satellite terminals,[5] sensory signals, and frequency division multiplexing (FDM) voice channels.[5] Although such signals, in principle, can be converted to digital form, the cost is at present often prohibitive. The major disadvantage of analog transmission is the increased susceptibility to noise degradation and nonlinearities within the transmission channel. A diagram of the essential elements of an analog transmission link[6] is shown in Figure 1. The model consists of three important elements: a transmitter, waveguide, and receiver. The transmitter usually contains an LED or laser diode as an optical source, and the electrical power driving the source is modulated in order to modulate the source output power directly. Thus, ideally, the source would transform the drive current waveform into an exact duplicate in output optical power.[7] Unfortunately, the sources have inherent nonlinearities so that a series expansion is necessary to represent the transmitter model:

$$P(t) = \sum_{j=0}^{J} K_j I^j(\omega_j t) \tag{1}$$

The above relationship is a series expansion containing a DC term ($j = 0$), followed by the fundamental term ($j = 1$) and all integer harmonics of the fundamental. For a perfectly linear device, $K_j = 0$ for $j \geq 2$, there are no harmonic components. The harmonic free case is expressed as the sum of an average term plus the signal.

$$P(t) = K_0 + K_1 I_1 (\omega_1 t) \tag{2}$$

An average term, not in the original signal, will be sensed by the photodetector. For instance, in baseband video transmission, the average term would result in a luminance error, and post detection signal processing may be needed to remove this error (e.g., clamping the sync tips in the video case).

The second block in Figure 1 represents the waveguide. Amplitude, phase, and group delay behavior with frequency become very important with analog transmission. Bandwidth, limited by modal distortion, is difficult to equalize since mode mixing is present and not readily predictable. A safer approach is to specify a fiber link with a flat response (amplitude and group delay) within the passband needed to carry the signal free of linear distortion.

The final functional block in Figure 1 is the receiver. The photodetector in the receiver converts the optical signal back into an electrical current that is a replica of the envelope of the optical power striking the detector. Detectors with bandwidth exceeding 1 GHz are available. This is adequate for most applications. The linearity of properly biased detectors is excellent and compares quite favorably with the linearity of optical sources. Thus the primary consideration for the receiver is its noise performance. This becomes especially important when several links are placed in tandem, since noise is cumulative in analog systems with repeaters.

The following sections in this chapter will provide more details on receiver sensitivity, systems with multiple channels, and multiple repeaters and some commonly used modulation techniques.

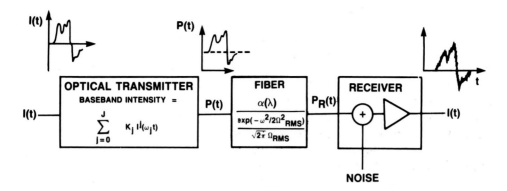

FIGURE 1.   Analog fiber optic link diagram.

II. CARRIER-TO-NOISE RATIO

The analysis of an analog transmission system must take into account: the type of signal[8] (i.e., modulation format and level), the number of multiplexed signals, the number of repeaters, and the system noise. These are the primary parameters common to all analog systems. Secondary parameters must include the characteristics of the individual components that form the transmitter, fiber, and receiver units. The analysis in most cases can be simplified by considering an individual repeater free link being driven by a simple sinusoid. A fairly straightforward calculation can be made to determine the rms carrier power to rms noise power at the optical receiver. The effects of multiple signals, multiple repeaters, and potential signal enhancement of the modulation technique modify the cnr calculation.

The signal generated at the transmitter is formed by forward biasing either an LED or laser in its most linear region. An example of laser transmitter operation is illustrated in Figure 2. The drive current through the laser is the sum of a direct current bias, I_o, and a time varying sinusoid representing anunmodulated carrier, $I_p Re\ e^{jwt}$

$$I_{drive} = I_0 + I_p\ Re\ e^{j\omega_c t} \tag{3}$$

The laser acts as a square law device so that the envelope of the output optical power has the same form as the input drive current, I_{drive}.

$$P_{out}(t) = \overline{P} + P_p\ e^{j\omega_c t} \tag{4}$$

It is important to realize that the above equation represents the envelope of an optical carrier(s). In a well-behaved system the optical carrier can be ignored and the system can be treated as a baseband system.[9] In fact, the optical carrier consists of a wide band of optical frequencies with random variation in frequency, phase, and amplitude. If one assumes a well-behaved system, the cnr calculation[6] is straightforward. The first important parameter is the intensity modulation index of the optical source. The intensity modulation index (m_i) is a measure of signal power at the receiver when the average received power is known (radiometers used in optical power measurements record average power). The definition of the modulation index is as follows when referring to Figure 2,

$$m_i = \frac{P_{PEAK}}{<P_T(t)>} \tag{5}$$

The cnr is defined as the ratio of rms signal power to the rms noise power. In the case of the sinusoidal signal received, the signal power is

$$C = \frac{1}{2}\left[\frac{\eta q}{h\nu} \, m_i \, \overline{P}_R <G>\right]^2 \qquad (6)$$

where m_i = intensity modulation index, $\overline{P}_R$ = average received power, η = quantum efficiency, q = the charge of an electron, h = Planck's constant, ν = optical carrier frequency (average), and G = detector gain (G = 1 for PIN detectors).

Several noise generators within the optical receiver[10] place a limit on the cnr. A fundamental limit is established by the quantum noise of the photodetector. Quantum noise is a condition caused by the statistical nature of photodetection and the particle nature of light. For a PIN-type photodetector, the quantum noise is a function of received power.

$$<i_Q^2> = 2q\left(\frac{\eta q}{h\nu}\right) \overline{P}_R \qquad (7)$$

where i_Q^2 = quantum noise spectral density, q = the charge of an electron, $\overline{P}_R$ = received average power, B_n = noise bandwidth, and $\eta q/h\nu$ = detector responsivity in amperes per watt.

Detectors may generate electron-hole pairs without incident radiation on the active area. This unwanted current is referred to as dark current, and it manifests itself as a noise source. Dark current creates shot noise that varies with applied detector bias voltage and temperature. The value of dark current noise due to surface generated carriers is proportional to the dark current amplitude.

$$<I_D^2> = 2qI_DB_N \qquad (8)$$

where $<I_D^2>$ = dark current noise spectral density, q = charge of an electron, I_D = dark current.

The noise generated within an avalanche photodiode (APD) is affected by the detectors mean gain $<G>$.[11] The quantum noise relationship expressed above is modified (increased) by an excess noise factor F($<G>$). The excess noise factor is a function of mean gain and a parameter k, which is a ratio of hole and electron ionization rates. The value of k extends between 0 and 1.0. The magnitude of k influences gain behavior as a function of electric field (indirectly the applied bias). The larger the k value, the greater the gain change in response to an electric field change. The excess noise factor is presented as a function of mean detector gain, $<G>$, and the ionization rate factor k.

$$F = k<G> + (2 - <G>^{-1})\,(1 - k) \qquad (9)$$

The value of k taken on by an APD depends on its fabrication and material. Germanium APD devices have a k factor close to unity, while silicon APDs fall between 0.02 and 0.04. As Equation 9 shows, the higher the k value, the greater the detector quantum noise for a fixed mean gain.

The noise generators introduced by the detector are signal-dependent quantum noise and signal-independent dark current noise. In addition to those noise sources created by the detector, the active stage in the detector preamplifier as well as physical resistors in the signal current path also create significant noise.[12] Physical resistors create thermal noise currents that are inversely proportional to the resistance value.

$$<i_T^2> = \frac{4kTB_N}{R} \qquad (10)$$

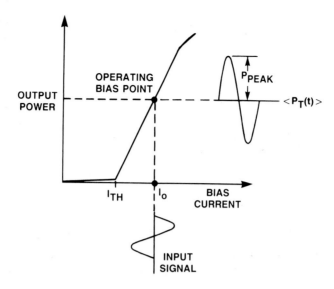

FIGURE 2. Direct intensity modulation for analog transmission.

where $<i_T^2>$ = thermal noise spectral density, k = Boltzmann's constant, T = absolute temperature, R = resistance, and B_N = noise bandwidth.

Finally, the active gain element itself contributes significant noise in an optical receiver. In a proper design, the first active stage noise dominates in the cnr calculations. Selection of the input stage device for the receiver's preamplifier is a critical decision in that it is one of the few options the receiver designer has available. The selection is based primarily on the needed signal bandwidth, the cnr specified, and the expected received power level. In general, both field effect devices and bipolar devices are used as the first gain stage. The amplifier configuration will also establish the receiver's operational behavior. Three available design approaches are (1) the transimpedance,[13] (2) high impedance, and (3) for very broadband applications a conventional 50 Ω gain block design. In all approaches, a circuit noise model is constructed to determine the amplifiers' noise contribution.

With a field effect transistor, FET, the device transconductance and the total capacitance from gate to ground (including: gate to source, gate to drain, and parasitic capacitances) are critical parameters in the noise model. The noise created by a FET is approximately equal to the expression below.

$$<i_{FET}^2> \simeq \frac{2kT\theta\omega^2C^2B_N}{g_m} \qquad (11)$$

where $<i_{FET}^2>$ = FET noise spectral density (grounded source), k = Boltzmann's constant, T = absolute temperature, ω = radian frequency, C = total capacitance at the gate port, g_m = FET transconductance, and θ = is a material-dependent constant (θ = −0.7 Si, θ = 1.1 GaAs).

The noise generated by the FET is dependent on frequency and a figure of merit equal to the C^2/g_m ratio.

The bipolar transistor has several noise sources that include thermal noise of the base-spreading resistance and shot noise created by the base and collector currents.

$$<i_{AMP}^2> = \left[4kTr_{bb} + \frac{2(kT)^2}{qI_c} + 2qI_c\left(\frac{1}{B}\left(r_{bb}^2+1\right) + \frac{f}{f_T}\left(r_{bb}^2+\frac{f}{f_T}\right)\right) \right]B_N \qquad (12)$$

where $<i^2_{AMP}>$ = the total device noise spectral density reflected to the input (grounded emitter), k = Boltzmann's constant, T = absolute temperature, r_{bb} = base spreading resistance, q = charge of an electron, I_C = collector current, B = device current gain small signal, f = frequency, f_T = device figure of merit.

The computation of cnr is straightforward once all noise terms have been quantified. The cnr specifically is the rms signal power divided by the rms noise power of all contributers. Assuming a sinusoid-shaped signal, the cnr is expressed below.

$$cnr = \frac{0.5\left[m_i\left(\frac{\eta 8}{h\nu}\right) \overline{P}<G> \right]^2}{(<i^2_Q> + <i^2_D> + <i^2_T> + <i^2_A>)B_N} \qquad (13)$$

where B_N is the noise bandwidth.

This expression assumes the use of an APD detector. For the PIN detector case, the mean gain term, $<G>$, equals unity as do the excess noise factors, F, in the quantum noise and bulk dark current noise terms.

It becomes an interesting exercise to examine how the cnr depends on relative levels of detector quantum noise and amplifier circuit noise. In most cases, it is safe to assume that dark currents are low enough not to significantly affect the cnr. With this simplification, the cnr is limited by detector quantum noise and thermal and shot noise of the detector preamplifier. Once the receiver design is set, the noise contribution of the detector preamplifier is constant. The detector quantum noise, however, is dependent on signal level. One can identify two distinct cases: the first occurs when amplifier noise dominates and the second when signal dependent quantum noise is dominant. In the former, the cnr can be increased by increasing the APD mean gain, $<G>$.

$$cnr\Big|_{\substack{receiver \\ noise \\ dominant}} = \frac{0.5\left[m_i P\left(\frac{\eta q}{h\nu}\right) <G> \right]^2}{[<i^2_T> + <i^2_A>]B_N} \qquad (14)$$

Since neither $<i^2_T>$ n or $<i^2_A>$ are affected by mean gain, $<G>$, the cnr increases at a rate proportional to mean gain squared, $<G>^2$. When the amplifier noise and quantum noise levels are in close proximity, the improvement decreases. In fact when the two noise source levels are equal or when quantum noise dominates, the cnr will decline. This brings us to the second case; the quantum noise level becomes the dominant noise source.

$$cnr\Big|_{\substack{quantum \\ noise \\ dominant}} = \frac{0.5\left[m_i\left(\frac{\eta q}{h\nu}\right) \overline{P}<G> \right]^2}{2q<G^2>\left(\frac{\eta q}{h\nu}\right) B_N} \qquad (15)$$

where m_i: source intensity modulation index, $0 <m_i\leq 1$, $\eta q/h\nu$: detector responsivity, q: charge of an electron, $<G^2>$: mean-squared gain, $<G^2> = <G>^2 F$, F: detector (APD) excess noise factor (see Equation 9), and B_N: noise bandwidth.

The above equation reveals that the cnr degrades inversely proportional to the APD photodetector's excess noise factor. Since the excess noise factor is an increasing function of APD gain, the cnr can be improved by reducing the APD gain. If for minimum APD gain the quantum noise is still dominant, a PIN diode might be a better detector choice. The correct detector selection for an analog system will depend primarily on the cnr specified

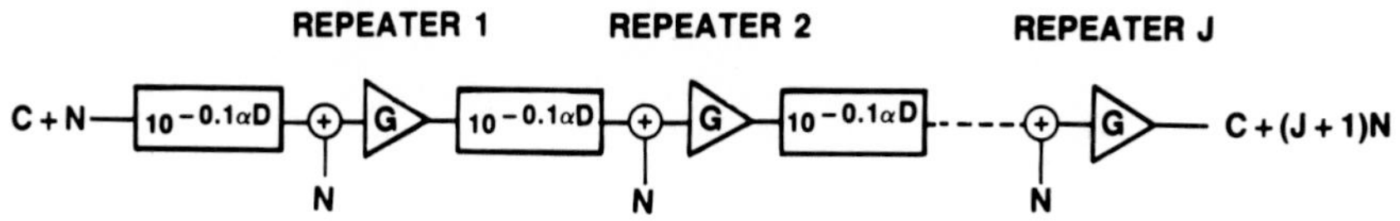

(1) $\left.\frac{C}{N}\right|_{OUTPUT} = (J+1)^{-1} \times \left.\frac{C}{N}\right|_{INPUT}$

(2) $G = 10^{\alpha D/10}$

(3) LINK INSERTION LOSS = $10^{-0.1\alpha D}$
WHERE α = FIBER LOSS IN dB/km
D = LINK DISTANCE IN km

(4) TRUNK INSERTION LOSS = $10^{-0.1\alpha DJ}$

FIGURE 3. CNR Model for an analog fiber optic link.

or the expected received power in an individual link. For the detector preamplifier noise-limited case, an optimum APD mean gain can be calculated (as discussed in the digital systems chapter) that results in a minimum received per level for a required cnr. In most cases, the cnr values needed for analog systems exceed 40 dB (10^4) that produces relatively low mean gain values ($\simeq$10). Such low gain figures are close to the lower operational limits of available APDs. Therefore, PIN diodes may be more common in actual system implementations. If this becomes the case, the cnr expressions are simplified because the mean gain in the expressions is set to unity, $\langle G \rangle = 1$, and excess noise is nonexistent.

III. MULTIPLE REPEATER SYSTEMS

Many applications will be encountered where a single optical fiber link can not meet the carrier-to-noise ratio requirement. This may be due to insufficient receiver sensitivity or excessive fiber insertion loss. In such cases, the total end-to-end distance must be divided into a number of fiber links, each of which terminates in a repeater.[14] The repeater consists of a receiver, a midsection gain stage, and ends with a linear optical source and modulator. If the ratio of output optical signal power to the input power is set equal to the total link insertion loss, the overall end-to-end system gain will be unity, and the signal power leaving the trunk equals the power entering the trunk. The configuration shown in Figure 3 represents a hypothetical trunk comprised of J optical links.

The end-to-end performance of the trunk is directly determined by each of the J links. A signal entering the trunk has a carrier power of C watts and accompanying noise power of N watts. Each repeater will add additional noise to the transmitted signal as the signal travels through each link. The signal also experiences loss introduced by the fiber, connectors, and splices. Each repeater compensates for the optical insertion loss so that the signal power leaving the trunk is the same power level that entered the system. The noise is also alternatively attenuated and amplified; however, the noise generated at each repeater accumulates. Thus the trunk can only degrade the cnr of the input signal, at best the degradation will be negligible. This can be achieved if the cnr of the trunk (cnr with an ideal noise-free input) exceeds the incoming signal's cnr by more than 10dB.

The noise introduced at the repeaters adds in a scalar fashion so that the cnr at the input and output (of the trunk) can be expressed as shown below.

$$\left.cnr\right|_{output} = J^{-1}\left(\left.cnr\right|_{input}\right) \qquad (16)$$

where J = number of repeaters including the end terminal.

It is interesting to consider the relationship between individual link loss and the total end-to-end loss of the complete trunk. The insertion loss of a link can be expressed in terms of fiber distance, D, and the fiber attenuation, α, in dB/km,

$$\text{Link insertion loss} = \alpha D, \text{ dB} \qquad (17)$$

The total trunk loss is the sum of the individual link losses. Trunk loss = JαD, dB where: J is the number of equal distance loss links, α is the fiber attenuation in dB/km, and D is the link distance in kilometers.

The cnr expression for an optical receiver shown earlier can be restructured to show a relationship between link loss (D) and trunk loss (αJD)

$$JK_o = \frac{0.5\left[m_i\left(\frac{\eta q}{h\nu}\right) \overline{P}_T <G>\right]^2 \alpha D \ 10^{-0.1\alpha D}}{[2q\overline{P}_T<G^2>10^{-0.1}\alpha D + N_A]B_N} \qquad (18)$$

where K_o is a fixed value of cnr, (JαD) is the trunk loss, (αD) is the link loss, m_i is intensity modulation index, ($\eta q/h\nu$) is the detector responsivity, $<G>$ is detector mean gain, $<G^2>$ is detector mean squared gain, J is the number of fiber links, q is electron charge, and B_N is the noise bandwidth.

The above expression shows that a maximum trunk loss (i.e., end-to-end distance) can be achieved with a unique link loss (i.e., repeater spacing). Link losses above and below the optimum result in shorter trunk distances as shown in Figure 4.

IV. MULTIPLE SIGNAL TRANSMISSION

The previous analysis was concerned with the rather simple case of the transmission of a single channel. In many applications one would like to transmit more than one channel. The trivial solution is of course to send the signals over separate fibers. A more interesting approach is to multiplex the signals in the electrical domain and use the multiplexed signal as the modulating waveform.[15,16] A third solution is to use several optical carriers in the same optical fiber; this is usually referred to as wavelength division multiplexing or WDM.[3]

In Figure 5 a number of baseband signals are assigned a set of carriers, ω_1, ω_2, . . . ω_m, where each carrier has a different frequency. The modulated carriers are electrically combined to form a composite signal that directly modulates a single optical source. The FDM terminal may be repeated N times as shown in Figure 5. Each optical source is assigned a different wavelength, and all of the modulated optical signals are combined so that all are launched into a single fiber waveguide as shown in Figure 6. Both FDM and WDM will extract a penalty on the transmission quality.[17] Electrical frequency division multiplexing (FDM) produces a composite modulation signal that drives a single optical source. Thus the detected signal power at the receiver is reduced in proportion to the number of signals multiplexed at this source. That is, the calculated cnr based on the assumption of a single sinusoid plus the expression $-10 \log_{10}M$, where M is the number of signals frequency multiplexed and used to modulate a single optical source. Figure 7 represents the cnr degradation factor for a number M multiplexed signals and J repeated links. The degradation factor assumes statistical averaging of the combined signals. This is a safe assumption for large values of M (combination of two or three channels may not average because the signals will add in voltage rather than power so that the degradations will be a 20 logM function).

Examination of the cnr curves of Figure 7 shows that system performance degrades rapidly as the number of FDM channels increases and repeaters are added to the system. This is an important limitation in the performance of analog fiber optic systems.

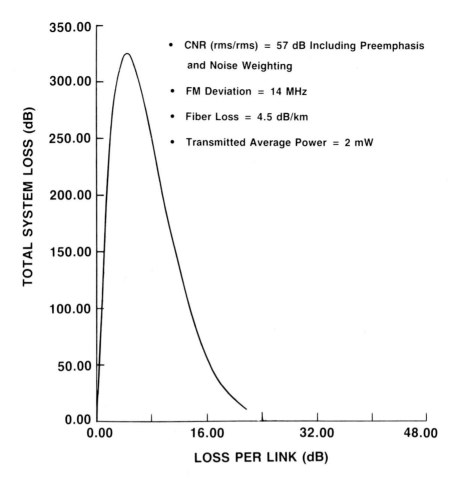

FIGURE 4. Total system (trunk) loss dependency on link (repeater) spacing.

The above analysis ignored nonlinearities in the transmission channel; the effect of such nonlinearities is to cause cross talk and intermodulation between frequency multiplexed channels and to degrade system performance beyond the performance predicted by the simple cnr analysis.[18-20] Careful selection of carrier frequencies is usually required to minimize cross talk and intermodulation.

V. MODULATION TECHNIQUES

In the previous sections, the tacit assumption was made that the information signal was directly modulated onto the optical carrier. An alternative technique is to use a subcarrier modulation technique. In this modulation scheme, the information signal modulates an ancillary subcarrier that is then intensity modulated onto the optical carrier. At the receiver the reverse process takes place, i.e., the modulated subcarrier is first recovered from the optical signal and then demodulated using conventional demodulation methods. The primary advantage of the subcarrier technique is that the ''demodulation gain'' obtained during demodulation of the subcarrier may improve performance as compared to direct intensity modulation. The subcarrier frequency and the bandwidth occupied by the modulation will affect the optical source selection and the required fiber and receiver bandwidth. The cnr of the subcarrier can be calculated using the equations derived previously. In order for a net improvement over direct intensity modulation to occur, the snr of the demodulated signal

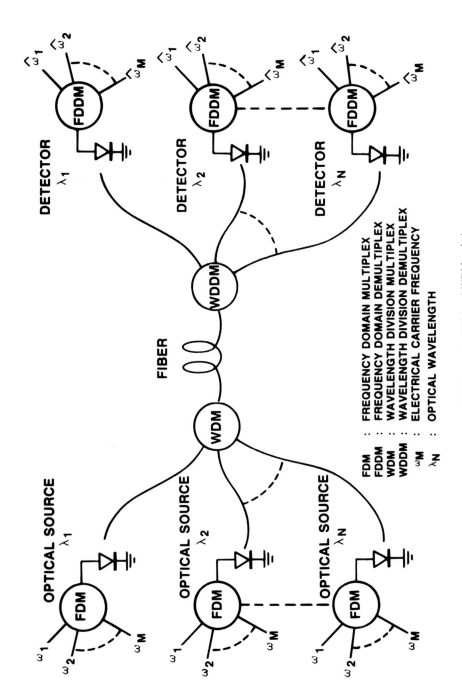

FIGURE 5.   Signal multiplexing using FDM and WDM techniques.

FDM   : FREQUENCY DOMAIN MULTIPLEX
FDDM   : FREQUENCY DOMAIN DEMULTIPLEX
WDM   : WAVELENGTH DIVISION MULTIPLEX
WDDM   : WAVELENGTH DIVISION DEMULTIPLEX
ω_M : ELECTRICAL CARRIER FREQUENCY
λ_N : OPTICAL WAVELENGTH

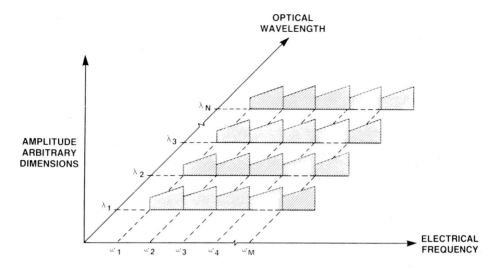

FIGURE 6. Frequency and wavelength signal assignments of a FDM and WDM system.

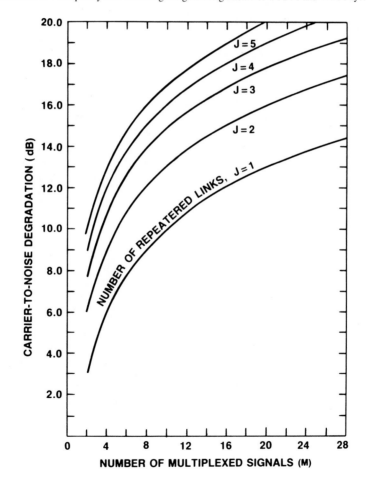

FIGURE 7. CNR degradation of frequency multiplexed signals with multiple repeaters.

must exceed the cnr of the subcarrier. This can be achieved with modulation methods that result in bandwidth expansion. AM modulation methods will not improve performance over direct intensity modulation, and standard AM will degrade performance by more than 3 dB.[21] Other bandwidth expansion technques that can be used for analog transmission are pulse position modulation (PPM) and pulse duration modulation (PDM).

A. Frequency Modulation

The use of subcarrier FM has gained wide acceptance, particularly in fiber transmission for CATV and microwave FDM voice systems. Frequency modulation will provide a post detection gain in snr over cnr provided that the cnr is above a threshold value. As long as the cnr exceeds the FM threshold limit, the optical receiver should require less power using FM than with baseband or AM signaling. The message signal undergoes integration prior to deviating the instantaneous frequency of the carrier signal.

$$A_c\cos\left[\omega_c t + \Delta\omega\int_0^t A_m\cos\omega_m\tau d\tau\right] \tag{19}$$

where A_c: peak carrier amplitude, A_m: peak message (modulation) amplitude, ω_c: carrier radian frequency, ω_m: message radian frequency, and τ: a dummy variable. Carrying out the integration results in the well-known simplification of the second term of the argument of the cosine,

$$\frac{A_m\Delta f}{f_m} \sin\omega_m t = \beta \sin\omega_m t \tag{20}$$

β is the maximum phase deviation produced by the modulating signal (of frequency ω_m) and is referred to as the FM modulation index. The FM advantage (improvement in post-demodulation snr over predemodulation cnr in the same bandwidth) increases as β increases, provided the FM threshold cnr has been exceeded. Increased transmission bandwidth is the price paid for the improved noise tolerance of the FM approach.* Exact amounts of improvement will also depend on preemphasis provided to the signal prior to frequency modulation. Preemphasis filtering uses the triangular noise variation ($\sim f^2$) characteristic to improve the snr value and to reduce the linearity requirements of the FM circuitry.

Most signals have much greater complexity than simple tone modulation. An example is the use of FM in the transmission of video. FM is a preferred modulation for video optical fiber links. Clayton[22] has developed a convenient relationship that predicts the FM advantage of a video FM signal

$$snr = \left(\frac{12\Delta F_s^2}{b_n^3}\right)(cnr) \tag{21}$$

where snr is the past demodulation signal-to-noise ratio as the result of FM advantage, and all baseband filter networks; ΔF_s is one half the peak to peak frequency deviation created by the peak white to blanking level of the video signal; and b_n is the noise bandwidth of all baseband filter networks including low pass weighting, preemphasis and deemphasis filters.

In the above relationship, the noise bandwidth b_n is generally set by existing standards. The one design option is the frequency deviation, ΔF_s. By doubling the FM deviation, a 6 dB improvement in snr is obtained for a fixed cnr level. The upper limit constraining the

---

* The relationship between FM deviation F_s and FM bandwidth f_b is described by Carson's rule as $f_b = 2 (\Delta F_s + 1)W$.

deviation is set by available FM bandwidth, threshold limit, and practical limits on deviator/discriminator linearity requirements and, of course, cost requirements.

B. Pulse Modulation

A useful modulation technique for fiber optic transmissions that overcomes the potential problems created by optical source nonlinearity is either pulse duration modulation (PDM) or pulse position modulation (PPM).[23] Such techniques straddle the line separating analog and digital transmission. The two techniques are related in that the information is implanted in the edge crossings of the modulation signal. The width of the Nth pulse of a PDM scheme is as shown below:

$$\sigma(N) = \sigma_0(1 + \gamma z(NT_{SA})) \tag{22}$$

where $\sigma(N)$ is the width of pulse N, σ_o is the shortest duration pulse, γ is the modulation depth index, and $Z(NT_{SA})$ is the pulse shape of the N pulse with sampling period T_{SA}. In PDM information is contained in the edge of the pulses and not in the pulses themselves. Thus narrow pulses contain as much information as the wider pulses and power is wasted in these wider pulses. In PPM, pulses of fixed duration are used with each pulse displaced from its nominal position.

The relationship between cnr and snr has been well covered in the literature for PDM and PPM techniques.[23,24] For average power the snr for PDM and PPM is shown below:

$$\left. snr \right|_{PDM} \leq \frac{<P>}{2} \left(\frac{f_b}{W}\right)(cnr) \tag{23}$$

$$\left. snr \right|_{PPM} \leq \frac{<P>}{8} \left(\frac{f_b}{W}\right)^2(cnr) \tag{24}$$

where P is average power, f_b is the pulse bandwidth, W is the message signal bandwidth.

Equation 23 shows that a net improvement is possible with PDM. The improvement is directly proportional to the carrier or pulse bandwidth (f_b). On the other hand, Equation 27 reveals a squared proportionality with carrier bandwidth (f_b). This is similar to the FM demodulation gain described in Equation 21. PPM is superior to PDM with regard to demodulation improvement. Because of the reduced pulse duty cycle PPM requires less average power than a CW FM signal, which may extend the lifetime of the optical source. However, FM provides a larger demodulation gain.

REFERENCES

1. **Fox, J., Fordham, D. I., Wood, R., and Ahern, D. J.,** Initial experience with the Milton Keynes optical fiber cable TV trial, *IEEE Trans. Commun.,* COM-30 (9), 2155, 1982.
2. **Buenning, H., Kreutzer, H. W., and Schmidt, F.,** Subscriber stations in service integrated optical broadband communication systems, *IEEE Trans. Commun.,* COM-30 (9), 2163, 1982.
3. **Asatani, K., Watanabe, R., Nosu, K., Matsumoto, T., and Nihei, F.,** A field trial of fiber optic subscriber loop systems utilizing wavelength division multiplexers, *IEEE Trans. Commun.,* COM-30 (9), 2172, 1982.
4. **Pan, J. J.,** 5 GHz Wideband Fiber-Optic Link, Technical Digest, Topical Mtg. Optical Fiber Communications, New Orleans, February 28 to March 2, 1983, pp. 74, 76.
5. **Albanese, A. and Lenzing, H. F.,** IF lightwave entrance links for satellite earth stations, *ICC'79 Conf. Rec.,* 1, 1.7.1, 1979.

6. **McDevitt, F. R., Hamilton-Piercy, N., and Hemmings, D. F.,** Optimized designs for fiber optic cable television systems, *IEEE Trans. Cable Television,* CATV-2, 169, 1977.

7. **Wenke, G. and Elze, G.,** Investigation of optical feedback effects on laser diodes in broad-band optical transmission systems, *J. Opt. Commun.,* 2, 128, 1981.

8. **Kanada, T., Hakoda, K., and Yoneda, E.,** SNR fluctuation and nonlinear distortion in PFM optical NTSC video transmission systems, *IEEE Trans. Commun.,* COM-30(8), 1868, 1982.

9. **Sato, K. and Asantani, K.,** Speckle, noise reduction in fiber optic analog video transmission using semiconductor laser diodes, *IEEE Trans. Commun.,* COM-30 (7), 1017, 1981.

10. **Smith, R. G. and Personick, S. D.,** Receiver design for optical fiber communication systems, *Top. Appl. Phys.,* 39, 89, 1982.

11. **Webb, P., McIntyre, R., and Conradi, J.,** Properties of avalanche photodiodes, *RCA Rev.,* 35, 244, 1974.

12. **Oliver, B. M.,** Thermal and quantum noise, *Proc. IEEE,* 53, 436, 1965.

13. **Hullett, J. and Moustakas, S.,** Optimum transimpedance broadband optical preamplifier design, *Opt. Quant. Electron.,* 13, 65, 1981.

14. **Cummings, D. O.,** Some considerations for amplifier spacing for AM and FM super trunks, *IEEE Trans. Cable Television,* CATV-1, 63, 1976.

15. **Baack, C., Elze, G., Grosskopf, G., Kraus, F., Krick, W., and Kiiller, L.,** Analog optical transmission of 26 TV channels, *IEEE Electron. Lett.,* 15(10), 300, 1979.

16. **Hara, E. H. and Ozeki, T.** Optical video transmission by FDM analogue modulation, *IEEE Trans. Cable Television,* CATV-2 (1), 19, 1977.

17. **Kersten, R. Th. and Rocks, M.,** Wavelength division multiplexing in optical communication systems, *J. Opt. Commun.,* 3, 93, 1982.

18. **Arnold, B.,** Third order intermodulation products in a CATV system, *IEEE Trans. Cable Television,* CATV-2 (2), 67, 1977.

19. **Sasaki, T. and Hataoka, H.,** Generation inter or cross modulation distortions in non-linear devices, *IEEE Trans. Cable Television,* CATV-2 (3), 131, 1977.

20. **Daly, J. C.,** Fiber optic intermodulation distortion, *IEEE Trans. Commun.,* COM-30(8), 1954, 1982.

21. **Peebles, P. Z.,** *Communications Systems Principles,* Addison-Wesley, Reading, Mass., 1976, chap. 5.

22. **Clayton, L.,** FM television signal-to-noise ratio, *IEEE Trans. Cable Television,* CATV-1 (1), 25, 1976.

23. **Carlson, A. B.,** *Communication Systems,* 2nd ed., McGraw-Hill, New York, 1975, 310.

24. **Schwartz, M., Bennett, W., and Stein, S.,** *Communication Systems and Techniques,* McGraw-Hill, New York, 1966, 248.

Chapter 8

IMAGING SYSTEM USING GRADIENT INDEX FIBERS

To R. Hsing

TABLE OF CONTENTS

I. INTRODUCTION

Fiber optics has become a well-defined technology. Much of the effort to date has been limited to optical communication applications, e.g., telephone companies are installing major links using optical fibers. However, interest in fiber optics imaging systems has been growing since the first light–focusing fiber rod with gradient index profile was made with glass materials in late 1968. The major characteristic of the gradient index (GRIN) fiber is that each single fiber rod is equivalent to a conventional lens and can be used to transmit an image. This kind of fiber — with its lens-like behavior — has begun to be widely used in imaging systems.

There are three types of gradient refractive indexes that have a focusing property: axial gradient, radial gradient, and spherical gradient. Gradient index work related to imagery started early in the 19th century when Maxwell studied the spherical gradient problems of the so-called Maxwell fisheye lens.[1] The Wood lens, which had cylindrical symmetry and a plane surface with a radial gradient, was invented around 1905.[2] The refractive index of the first fiber rod that had imaging property was also fabricated with a radially gradient distribution. It is of interest that these gradients are also found in human eyes and in the eyes of some animals.[3] Most recently, this kind of fiber has brought broad attention to the applications of imaging systems, coupling lens, and microoptical circuits. At least three companies have incorporated fiber optics into their copier/duplicator products. We will concentrate these discussions on the radial gradient type.

This chapter will focus on discussions of fiber imaging systems. It will have, after this introduction, sections on: II. Imaging Theory; III. Measurement Techniques; IV. Fabrications; and V. Applications. The presentation will be general, which we hope, will stimulate interest in the state-of-the-art of fiber optics imaging systems.

II. IMAGING THEORY

The index of refraction is homogeneous in a conventional lens, i.e., it is constant through the entire material. The lens designer can change the lens curvature to achieve focusing properties.

As shown in Figure 1a, let the focal length of a conventional thin lens be f. The relationship of a complex field amplitude of an incident optical field between the immediate right of the lens, $E_R(\vec{r})$, and the immediate left, $E_L(\vec{r})$, is[4]

$$E_R(\vec{r}) = E_L(\vec{r})\exp\left(-ik\frac{r^2}{2f}\right) \tag{1}$$

where r is the position vector of the measured point that is the radial distance from the optical axis, $k = 2\pi n/\lambda_o$ is the wave vector, and n is the refractive index of the lens. Phase shift, $kr^2/2f$, is caused by the lens effect and increases quadratically with the radial distance from the center axis.

In a similar case, as shown in Figure 1b, it is possible to fabricate a fiber lens element that has flat faces at both ends and whose refractive index is distributed in the radial direction with a parabolic form inside the material, i.e.,

$$n(r) = n(o)\left(1 - \frac{K_2}{2k}r^2\right) \tag{2}$$

where K_2 is a constant. The phase delay of a wave propagating through a section, dz, of an inhomogeneous medium with the refractive index, n, is

$$\varphi = kdz = \frac{2\pi}{\lambda_o} n(o)\left(1 - \frac{K_2}{2k} r^2\right)dz \qquad (2a)$$

It is clear that a thin lens element with the refractive index described by Equation 2 will also have lens behavior.

A. Ray Optics and Image Formation Formulas

Figure 2 shows a light ray propagating along the optical axis, z, through a lens medium. Weak focusing with the refractive index of parabolic distribution is assumed. The meridional rays of the first order Gaussian optics are considered here. The refractive index for this parabolic type needed for proper imaging is[5]

$$n(r) = n(o)\left(1 - \frac{A}{2} r^2\right) \qquad Ar^2 << 1 \qquad (3)$$

where $n(o)$ is the refractive index along the optical axis. A is a positive constant that represents the gradient of the refractive index.

The behavior of the light rays propagating through an optically inhomogeneous medium can be described as follows:[6]

$$\frac{d}{ds}\left(n \frac{d\vec{r}}{ds}\right) = \nabla n \qquad (4)$$

where s is the distance along the ray, and r is the position vector of the point at s. For a paraxial ray in the one-dimensional case, Equation 4 can be expressed as

$$\frac{d^2r}{dz^2} + Ar = 0 \qquad (5)$$

Applying the initial condition at the input plane, $z = 0$, that the light ray has radius γ_i and slope γ_i', the solution of Equation 5 is

$$\gamma(z) = \cos(\sqrt{A}\, z)\gamma_i + [1/\sqrt{A}]\sin(\sqrt{A}\, z)\gamma_i'$$
$$\gamma'(z) = -\sqrt{A}\,\sin(\sqrt{A}\, z)\gamma_i + \cos(\sqrt{A}\, z)\gamma_i' \qquad (6)$$

The sinusoidal ray paths with a period of $2\pi/\sqrt{A}$ were observed from Equation 6. From Snell's law:

$$r'_{o_{air}} = n(o)r'_o$$
$$r'_{i_{air}} = n(o)r'_i \qquad (6a)$$

The relationship of the ray position and the corresponding slope between the input plane at $z = 0$ and the output plane at $z = 1$ of a fiber rod can be expressed by a matrix form:

$$\begin{vmatrix} r_o \\ \\ r'_{o_{air}} \end{vmatrix} = \begin{vmatrix} \cos(\sqrt{A}\, 1) & [n(o)\sqrt{A}]^{-1}\sin(\sqrt{A}\, 1) \\ \\ -\sqrt{A}\, n(o)\sin(\sqrt{A}\, 1) & \cos(\sqrt{A}\, 1) \end{vmatrix} \begin{vmatrix} r_i \\ \\ r'_{i_{air}} \end{vmatrix} \qquad (7)$$

In the paraxial case, the position and its slope at the entrance and exit planes of an optical system can always be represented by the above matrix, a so-called ABCD or beam matrix. Several interesting results can be derived from this matrix. Now, simplifying Equation 7:

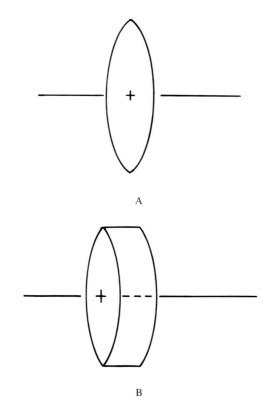

A

B

FIGURE 1. (a) The conventional lens with curved faces and homogeneous index; and (b) the fiber rod lens with flat-faces and gradient index.

$$\begin{vmatrix} r_o \\ r'_{o_{air}} \end{vmatrix} = \begin{vmatrix} A & B \\ C & D \end{vmatrix} \begin{vmatrix} r_i \\ r'_{i_{air}} \end{vmatrix} \tag{7a}$$

and AD-BC = 1 is always true according to the principle of light reversibility.

As shown in Figure 3, let the distance between the entrance point of the incident light and the first principle plane be h_o. The focal point on the optical axis in the entrance plane is S_o. The following simple relations are easily derived:

$$f = S_o + h_o$$

$$S_o = -\left.\frac{r_i}{r'_i}\right|_{r'_o=0} = \frac{\cos(\sqrt{A}\ 1)}{n(o)\sqrt{A}\ \sin(\sqrt{A}\ 1)}$$

$$h_o = \left.\frac{r_o - r_i}{r'_i}\right|_{r'_o=0} = \frac{1}{n(o)\sqrt{A}}\ \tan\left(\frac{\sqrt{A}\ 1}{2}\right) \tag{8}$$

and then

$$f = \frac{1}{n(o)\sqrt{A}\ \sin(\sqrt{A}\ 1)} \tag{9}$$

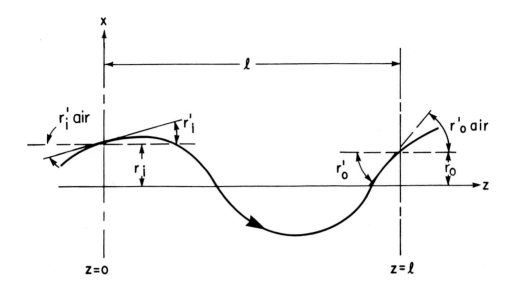

FIGURE 2. Light path in the lens-like medium

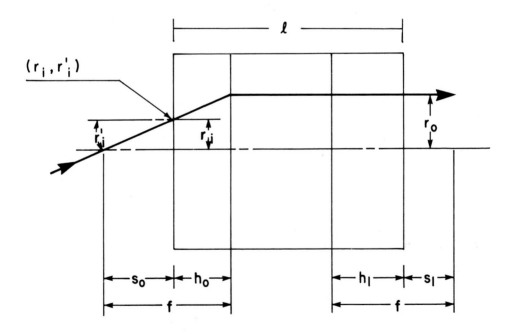

FIGURE 3. Focal length and principal plane of rod-lens

Similarly, from Figure 4 and the ABCD matrix, one can obtain the following matrix form:

$$
\begin{vmatrix} r_o \\ r_o' \end{vmatrix} = \begin{vmatrix} 1 & 1_1 \\ 0 & 1 \end{vmatrix} \begin{vmatrix} \cos(\sqrt{A}\ 1) & [n(o)\sqrt{A}]^{-1}\sin(\sqrt{A}\ 1) \\ -\sqrt{A}\ n(o)\sin(\sqrt{A}\ 1) & \cos(\sqrt{A}\ 1) \end{vmatrix} \begin{vmatrix} 1 & 1_o \\ 0 & 1 \end{vmatrix} \begin{vmatrix} r_i \\ r_i' \end{vmatrix}
$$

Then:

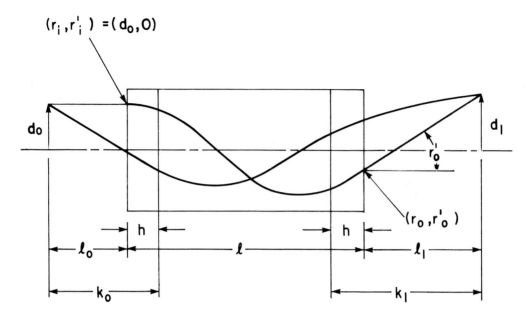

FIGURE 4. Imaging rules by a GRIN-rod lens

$$r_o = [\cos(\sqrt{A}\ 1) - 1_i n(o)\sqrt{A}\ \sin(\sqrt{A}\ 1)]r_i$$
$$+ \{1_o\cos(\sqrt{A}\ 1) - 1_o 1_i n(o)\sqrt{A}\ \sin(\sqrt{A}\ 1)$$
$$+ [n(o)\sqrt{A}]^{-1}\sin(\sqrt{A}\ 1) + 1_i\cos(\sqrt{A}\ 1)\}r_i' \tag{10}$$

Since the object point r_i is focused at image point r_o the second term in Equation 10 should vanish, i.e.,

$$1_o\cos(\sqrt{A}\ 1) - 1_o 1_i n(o)\sqrt{A}\ \sin(\sqrt{A}\ 1)$$
$$+ [n(o)\sqrt{A}]^{-1}\sin(\sqrt{A}\ 1) + 1_i\cos(\sqrt{A}\ 1) = 0$$

The relationship between the object distance 1_o and the image distance 1_i is

$$1_i = \frac{1}{n(o)\sqrt{A}}\left[\frac{\sin(\sqrt{A}\ 1) + n(o)\sqrt{A}\ 1_o\cos(\sqrt{A}\ 1)}{n(o)\sqrt{A}\ 1_o\sin(\sqrt{A}\ 1) - \cos(\sqrt{A}\ 1)}\right] \tag{11}$$

The magnification $m = d_i/d_o$, and $d_i = 1_i r_o' + r_o$

$$\begin{vmatrix} r_o \\ r_o' \end{vmatrix} = \begin{vmatrix} \cos(\sqrt{A}\ 1) & [n(o)\sqrt{A}\ 1)]^{-1}\sin(\sqrt{A}\ 1) \\ -\sqrt{A}\ n(0)\sin(\sqrt{A}\ 1) & \cos(\sqrt{A}\ 1) \end{vmatrix} \begin{vmatrix} r_i \\ 0 \end{vmatrix}$$

Then $m = (1_i r_o' + r_o)/r_i$

$$= -1_i n(o)\sqrt{A}\ \sin(\sqrt{A}\ 1) + \cos(\sqrt{A}\ 1)$$
$$= -[n(o)\sqrt{A}\ 1_o\sin(\sqrt{A}\ 1) - \cos(\sqrt{A}\ 1)]^{-1} \tag{12}$$

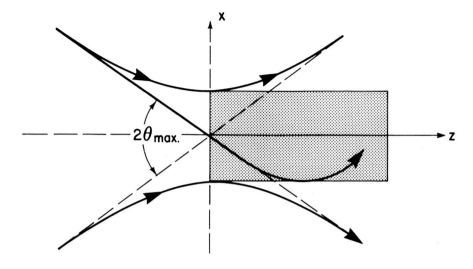

FIGURE 5.   Field angle of a GRIN-rod lens

An erect image is obtained if m > 0, and an inverted image if m < 0. One cannot obtain an erect image with a conventional singlet spherical lens. Because of this characteristic, it is possible to use a lens array consisting of many individual fiber rods in a multiaperture system.

The light collecting capability of traditional fibers can be quantified by their numerical apertures (NA) and the half-angles of the incident light. The amount of light flux captured by a gradient index fiber rod is determined by its field angle. This is similar to the conventional optical fiber where this angle depends on fiber radius. It has maximum value at the center and decreases to zero at the periphery. As shown in Figure 5, the ray trajectory is a hyperbola whose asymptote is the maximum half-field angle at the optical axis, and the vertex is R. Therefore, it can be expressed as

$$\frac{x^2}{R^2} - \frac{z^2}{R^2\cot^2\theta} = 1 \tag{13}$$

Let this maximum field angle be θ_{max}. One can have:

$$\tan\theta_{max} = \sqrt{2n(0)[n(0) - n(R)]} \tag{14}$$

In fact, Equation 14 is very similar to the NA of an ordinary conventional optical fiber:

$$NA = \sqrt{n^2(o) - n^2(R)} \approx \sqrt{2n(o)[n(o) - n(R)]} \tag{15}$$

A more general representation of the index distribution of a GRIN-rod lens can be written as[7]

$$n^2(r) = n^2(o)[1 - (\sqrt{A}\, r)^2 + h_4(\sqrt{A}\, r)^4 - h_6(\sqrt{A}\, r)^6 + \ldots\,] \tag{16}$$

where n(o) is the index at the center axis. Physical and geometrical optics[9] show that all

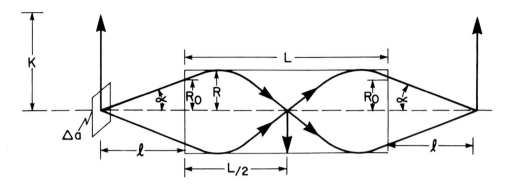

FIGURE 6.   Side view of a GRIN fiber. For a Lambertian source Δa on the optical axis, all image-forming rays are within the cone defined by the angle α. (From Rees, J. D. and Lama, W., *Appl. Opt.*, 19, 1065, 1980. With permission.)

meridional rays can be imaged precisely (geometrically) if the refractive index has the following optimum distribution:

$$n^2(r) = n^2(o)\text{sech}^2(\sqrt{A}\ r)$$
$$= n^2(o)[1 - (\sqrt{A}\ r)^2$$
$$+ (2/3)(\sqrt{A}\ r)^4 - (17/45)(\sqrt{A}\ r)^6 + \ldots] \tag{17}$$

For skew rays, the ideal index distribution is of the form:[10]

$$n^2(r) = n^2(o)[1 + (\sqrt{A}\ r)^2]^{-1}$$
$$= n^2(o)[1 - (\sqrt{A}\ r)^2 + (\sqrt{A}\ r)^4 - (\sqrt{A}\ r)^6 + \ldots] \tag{18}$$

but, it is difficult to make the kind of distribution shown in Equation 18. It is clear that there is no optimum distribution for all rays. For a weak focusing lens ($\sqrt{A}r << 1$) of Equation 3, the index distribution is the first-order expansion of the above equations.

B. Radiometric Properties

The radiometric properties are among the important parameters that imaging system engineers should consider. The exposure and the radiometry of a fiber array and an individual fiber rod have been analyzed by Rees and Lama.[11] An erect image formed by a weak focusing fiber with unit magnification of an individual GRIN rod was assumed in Figure 6. The on-axial irradiance for the image is

$$E_o = \pi NT\sin^2(\alpha) \tag{19}$$

where N is the Lambertian radiance, T is the fiber transmission, α is the maximum aperture angle. Similar results had been observed for conventional homogeneous lenses whose exit pupils subtend the angle 2α. The off-axis irradiance from a single fiber E(x,y) was calculated by Matsushita and Toyama[12] and follows an ellipsoial distribution:

$$\left(\frac{E}{E_o}\right)^2 + \left(\frac{x}{K}\right)^2 + \left(\frac{y}{K}\right)^2 = 1$$
$$x^2 + y^2 \leq K^2 \tag{20}$$

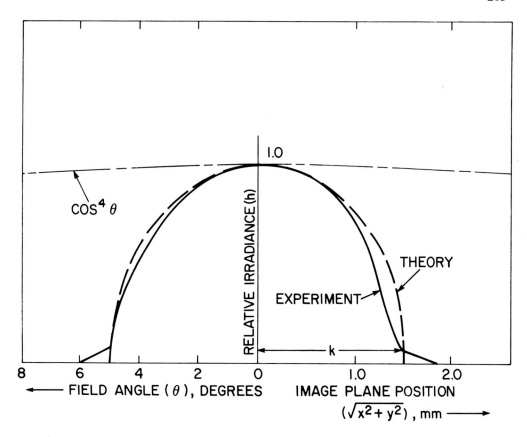

FIGURE 7. Relative irradiance distribution in the image plane or a single GRIN fiber. The right abscissa is the position in the image plane measured from the optical axis. The left abscissa is the corresponding field angle θ. Theoretically, the irradiance distribution is ellipsoidal [Equation 20]. For this fiber, R = 0.48 mm, L = 31.0 mm, and $(A)^{1/2} = 0.122$ mm^{-1}. These values yield a maximum field height k = 1.52 mm, in good agreement with the measured value. For comparison, the cos^4 θ falloff of a conventional lens is also plotted. (From Rees, J. D. and Lama, W., *Appl. Opt.*, 19, 1065, 1980. With permission.)

where K is the maximum field height (see Figure 6). Good agreement between experimental data[11] and theory is shown by Figure 7. One can see that the irradiance distribution at the image plane falls off more rapidly than the cos^4 law for a conventional lens.

Many GRIN rods can be packaged together to form a fiber array. Compared to a conventional spherical lens, it has a shorter focal length, shorter conjugate length, and numerous apertures. Figure 8 shows that good uniformity is observed[11] along the linear array direction with ±3% modulation. This is an important feature for information processing applications, such as copier/duplicator and facsimile. The high uniformity comes from the compensation of the NA of each individual fiber rod. Another feature of a fiber array is that the average exposure is independent of the overall conjugate length.

C. Modulation Transfer Function (MTF)

The resolution capability of an imaging system is determined by its modulation transfer function (MTF). The requirement of a lens MTF for a copier/duplicator is 45% of 5 lp/mm or more and 20% of 10 lp/mm or higher with a white illumination over the entire image plane.[13] Figure 9 is an interesting comparison of the MTF of a fiber array and a conventional lens for an 85-mm image height. As can be seen, the MTF is uniform within the entire image plane of the fiber array. However, the MTF is fully dependent upon the image height in the case of the conventional lens.

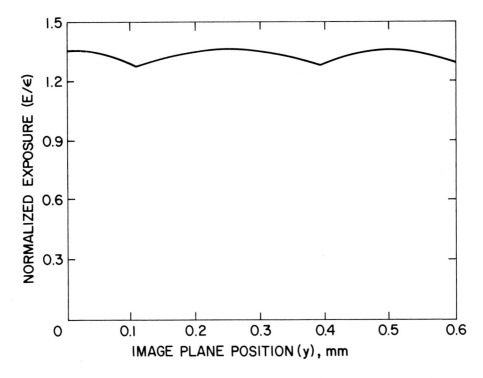

FIGURE 8.   Calculated normalized image plane exposure E/ε vs. distance y along the rows of a two-row SELFOC® array. The good spatial uniformity (modulation = ± 3%) is observed in this case. (From Rees, J. D. and Lama, W., *Appl. Opt,* 19, 1065, 1980. With permission.)

Figure 10, which is based on the theoretical work done by Levi and Austing,[14] shows the theoretical on-axis diffraction-limited MTF of a commercially available GRIN rod which is 1 mm in radius and 29.26 mm long under 550 nm wavelength illumination and with unit magnification.[15] Compared with the linear array, an individual fiber lens has a much higher MTF value. One of the major reasons for obtaining lower MTF in the linear array case is that the multiple apertures contribute aberrations to each of the images formed.

D. Depth of Focus

The depth of focus can be expressed by[12]

$$\Delta1 = \pm\ d\theta/\theta \tag{21}$$

where θ is the maximum half field viewing angle of the fiber rod, and dθ is the width of the resolution bar targets. For example, if the θ value is assumed to be 6° (≈0.1 rad), dθ = 0.125 mm for a 4 c/mm bar target. From Equation 21, the calculated value for the depth of focus is ±1.25 mm.

E. Aberration

A diffraction-limited optical system can image a point object precisely. Limitations other than the diffraction effect also cause distortions; these are called "aberrations". For most paraxial rays, usually the first step in designing optical systems is to layout the first and third order prototype systems.

The classical third-order aberration theories of inhomogeneous lenses with cylindrical index distributions were studied by Moore[16] and Sands[17] in early 1970. They were then extended to the fifth-order by Gupta et al. in 1976.[18] However, the analytic expressions for

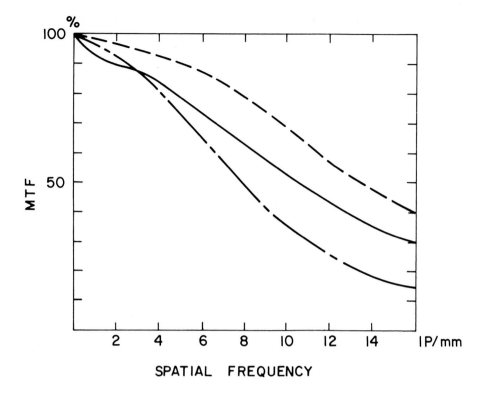

FIGURE 9. MTF of fiber array and lens; —, fiber array; — — — , copying lens center; —·— , copying lens image height 85 mm. (From Kawazu, M. and Ogura, Y., *Appl. Opt.*, 19(7), 1105, 1980. With permission.)

ray paths and the aberration theories developed by Streifer and Paxton[19] are more accessible to those who are concerned with the design of fiber optics imaging systems.

Using the same equations and procedures as Rawson et al.[10], Tomlinson[20] displayed spot diagrams of a GRIN rod lens with typical values of L = 32 mm, n(0) = 1.5 and NA = 0.2. This paper showed three different kinds of index distribution, but we will only discuss the one we feel is popular for imaging. Both the parabolic and sech distributions are adapted for calculations (see Figures 11 and 12). The values of r_a in these two figures are absolute distances measured from the optical axes. All the figures in Figure 11 are circularly symmetric, so no coma or astigmatism occurs for a GRIN rod lens with parabolic index distribution. Note also that all figures are centered, so there is no field distortion for this kind of index distribution. Figure 12 shows the spot diagrams of a GRIN rod lens with sech distribution. When the lens is a half-pitch long (i.e., z = 16 mm in this example), one can obtain a perfect image on the optical axis. The horizontal line of focus due to the astigmatism can also be observed in the off-axis focal plane. One can obtain smaller spot sizes at the other focal planes. The amount of field curvature for the sech distribution is about one third of that for the parabolic distribution.

It is also worth while to know the sensitivity between higher order aberrations and imaging characteristics. Gupta et al.[18] gave explicit expressions for fifth-order aberrations of rotationally symmetric optical systems. They also gave formulas for the fifth-order aberration coefficients of gradient index media. The Gupta et al. analysis also separated the contributions due to different kinds of aberration. Generally, the effect of higher order aberrations is increased rapidly with increasing NA and off-axis distance.

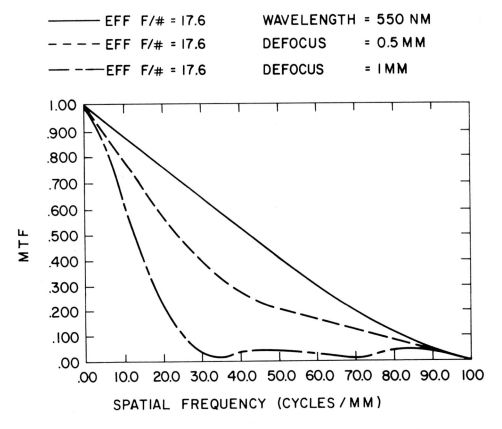

FIGURE 10.   The theoretical calculation on diffraction limited MTF of a GRIN rod lens. (From Seachman, N., personal communication, 1981. With permission.)

F. Spectral Transmittance

The spectral transmittance is also a useful parameter for applications on imaging systems, as, for example, in copying machines. The thallium-based fiber has the same spectral transmittance as the copying lens, but the cesium-based fiber is shifted to a shorter wavelength side. Basically, there is no substantial difference between the fiber lens and the glass lens.

G. Chromatic Aberration

One of the major applications for linear GRIN arrays is photocopiers in which white light or a light source with a broad spectrum are often used. Because of the dependence of the refractive index, n(r), and the gradient constant, A, on the wavelength, the chromatic aberration will degrade the image quality. Assuming that the refractive index has the form of Equation 3, the estimating function of chromatic aberration is expressed as[21]

$$\frac{\Delta L}{L} = \frac{1}{2} \frac{\dfrac{1}{Vr_o}\left(1 - \dfrac{1}{n_{r_o}}\right) - \left(\dfrac{1}{V(o)}\right)\left(1 - \dfrac{1}{n(o)}\right)}{\left(\dfrac{n(o)}{n_{r_o}}\right) - 1} \qquad (22)$$

where

$$V(o) = -\frac{n(o) - 1}{\Delta n_o}$$

$$n(r) = n_{(o)}[1 - (2\pi r/L)^2]^{1/2}$$
$$L = 32.0 \quad n_{(o)} = 1.5 \quad N.A. = 0.2$$

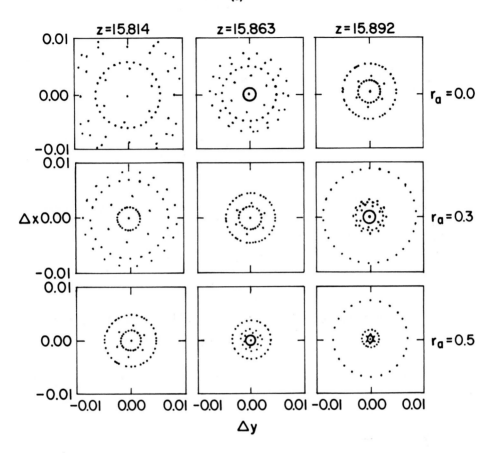

FIGURE 11. Calculated spot diagrams for a GRIN-rod lens with a parabolic refractive-index distribution. The assumed lens parameters are L = 32, n(o) = 1.5. The input rays are approximately uniformly distributed over a cone with NA = 0.2. (From Tomlinson, W. J., *Appl. Opt.*, 19(7), 1117, 1980. With permission.)

$$Vr_o = -\frac{n_{r_o} - 1}{\Delta n_{r_o}}$$

$$\Delta L = L_C - L_F$$

$$L = L_D$$

V(o) and Vr_o are the Abbe's numbers on the axis and periphery of a fiber rod. L_C, L_D, and L_F are the periodic lengths, respectively, for the hydrogen C line (0.6563 μm), the sodium D line (0.5893 μm), and the hydrogen F line (0.4861 μm). The parameters of n(o), n_{r_o}, V(o), Vr_o are the general specifications for a traditional lens system.

Figure 13 shows the relationships between $\Delta L/L$ and $\Delta V/Vr_o$ as functions of n(o), Vr_o, and $\Delta n = n(o) - n(a)$, where ΔV is the difference of Abbe's numbers at the optical axis and the periphery. The smaller value of $\Delta V/Vr_o$ that indicates a low chromatic aberration is desired.

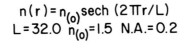

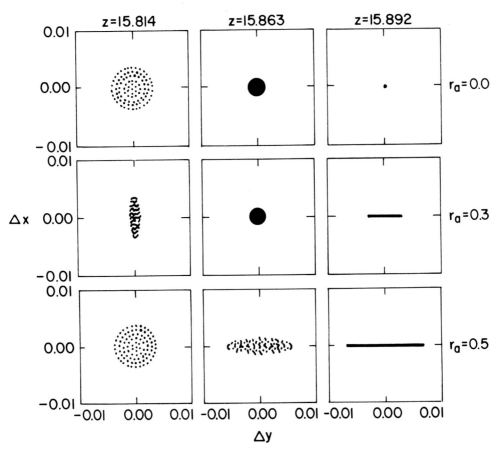

FIGURE 12.   Calculated spot diagrams for a GRIN-rod lens with a sech refractive-index distribution. The other parameters are the same as for Figure 11. (From Tomlinson, W. J., *Appl. Opt.*, 19(7), 1117, 1980. With permission.)

H. f-Number (f/No.)

In practical optical system design, the brightness of a lens is defined by the f/No. For a conventional spherical lens, the relationship between the illumination, I, and the f/No., F is

$$I = \frac{\pi K}{4} \left(\frac{1}{F}\right)^2 \frac{1}{(1 + m)^2} \tag{23}$$

where m is the magnification. For unit magnification, Equation 23 can be simplified to

$$I = \frac{\pi K}{16} \left(\frac{1}{F}\right)^2 \tag{24}$$

Since the gradient index linear array can only be used for unit magnification, the average illumination, I_{av}, within the slit width, w, of an array lens has been derived to be[13]

$$I_{av} = \left(\frac{\pi^4 K}{6}\right) \left(\frac{N}{w}\right) \left(\frac{D^3 n^2(o)}{p^2}\right) \tag{25}$$

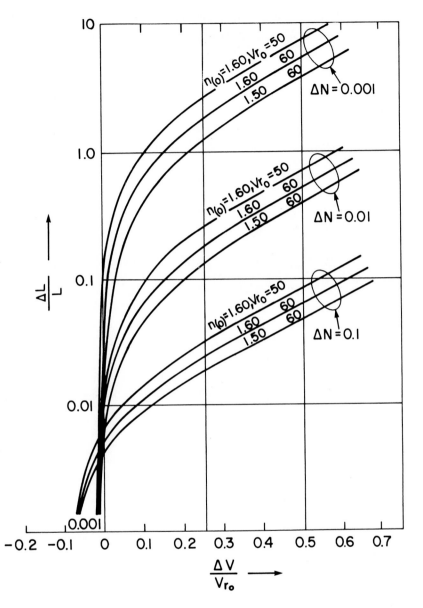

FIGURE 13. Relationship between $\Delta L/L$ and $\Delta V/V_{ro}$ as parameters of n(o), V_{ro}, and ΔN. (From Nishizawa, K., *Appl. Opt.*, 19(7), 1052, 1980. With permission.)

where N is the number of rows in the array, D is the diameter of each individual fiber rod, and p is the period of the sinusoidal ray path. Assuming that both Equations 24 and 25 are the same, the definition of f/No. was found as

$$F = \left[\left(\frac{16}{6} \right) \left(\frac{\pi^3 N}{w} \right) \left(\frac{D^3 n^2(o)}{p^2} \right) \right]^{-1/2} \tag{26}$$

The slit width, w, can be calculated as follows:

$$w = 2L\theta + (N - 1)\sqrt{3} \tag{27}$$

where L is the object distance, and θ the maximum acceptance angle. For a typical commercially available fiber array lens,[22] D $\approx$ 1.065 mm $\pm$ 0.015 mm, n(0) = 1.537, P $\approx$ 49.5 mm $\pm$ 1.5 mm, and θ = 6°. In this example, the f/No. was found to be around 5.

I. Wave Optics[23]

Geometrical optics can provide enough knowledge to allow designing an imaging system with reasonably good performance. However, it would be helpful to analyze the performance of the imaging system by wave optics.

As shown previously, the general expression of a refractive index distribution having the focusing property is

$$n^2(r) = n^2(0)[1 - (\sqrt{A}\, r)^2 + h_4(\sqrt{A}\, r)^4 - h_6(\sqrt{A}\, r)^6 + \ldots] \tag{28}$$

The distribution of the transverse field component is

$$E(r,z) = \Psi(r,z)\exp(j\omega t - j\beta z) \tag{29}$$

Substituting Equation 29 into the wave equation, the function Ψ for the index distribution of Equation 28 is given by the following equation in cylindrical coordinates:

$$\frac{1}{r}\frac{d}{dr}\left(r\frac{d\Psi}{dr}\right) + \frac{d^2\Psi}{d\theta^2} - \beta^2\Psi = -\frac{\omega^2}{c^2}\Psi \tag{30}$$

The characteristic spot size ω_o of the fundamental mode is defined as

$$\omega_o = \frac{a}{\sqrt{V}} \tag{31}$$

where V is the normalized frequency of the fiber, and

$$V = k_o\, n(0)\, a\sqrt{2\Delta} \tag{32}$$

where $k_o = 2\pi/\lambda$, a is the radius of a GRIN rod and n(0) is the refractive index at the optical axis;

$$\Delta = \frac{n(0) - n(a)}{n(0)} \tag{33}$$

The sinusoidal ray path will be perturbed if the value of ω_o/a is not small. For example, in a general GRIN rod lens, the value of ω_o/a is around 0.0137 for λ = 0.5 μm, n(0) = 1.5, a = 0.9 mm, and Δ = 5%. Usually, the value of ω_o/a is a indicator to tell us whether we can use this kind of a GRIN rod as an imaging lens. If the parabolic distribution of the refractive index was assumed (i.e., only the first two terms of Equation 28 are considered), the propagation constant β_{nm} associated with it is given by:[24]

$$\beta_{nm} = k(0) - (2n + m + 1)\sqrt{A} \tag{34}$$

where k(0) = $(2\pi/\lambda)$n(0), and n and m are the radial and azimuthal mode numbers. However, if we considered the more generalized expression of the refractive index as shown in Equation 18, the propagation constant, β_{nm}, is obtained by the perturbation method[7] as follows:

$$\beta_{nm} = k(0) - (2n + m + 1)\left[\frac{\sqrt{A}}{k(o)}\right]k(o)$$

$$+ \frac{1}{2}\left\{h_4\left[\frac{3}{2}(2n + m + 1)^2 + \frac{1}{2}(1 - m^2)\right]\right.$$

$$\left. - (2n + m + 1)^2\right\}\left[\frac{\sqrt{A}}{k(o)}\right]^2 k(o) + O\left\{\left[\frac{\sqrt{A}}{k(o)}\right]^3\right\}k(o) \qquad (35)$$

The propagation constant of the meridional rays that have a radially symmetric mode can be obtained easily by letting m = 0 in the above equation, i.e.,

$$\beta_{n0} = k(o) - (2n + 1)\left[\frac{\sqrt{A}}{k(o)}\right]k(o)$$

$$+ \frac{1}{2}\left\{h_4\left[\frac{3}{2}(2n + 1)^2 + \frac{1}{2}\right] - (2n + 1)^2\right\}\left[\frac{\sqrt{A}}{k(o)}\right]^2 k(o)$$

$$+ O\left\{\left[\frac{\sqrt{A}}{k(o)}\right]^3\right\}k(o) \qquad (36)$$

The β_{nm} term plays an important part in the fiber imaging system. The nonlinear terms of m and n in Equation 35 correspond to the wave aberration. The wave aberration amount is indicated by the ratio, R, of the third to the second term in Equation 25. By neglecting the $(1 - m^2)$ term, and letting L = 2n + m, the value R is given:

$$R = \frac{\left[\frac{\sqrt{A}}{k(0)}\right]^2 \frac{1}{2}\left\{(L + 1)^2\left(\frac{3}{2}h_4 - 1\right)\right\}}{(L + 1)\left[\frac{\sqrt{A}}{k(0)}\right]}$$

$$\approx \frac{3}{4}\left[\frac{\sqrt{A}}{k(0)}\right]\left(h_4 - \frac{2}{3}\right)(L + 1)$$

For most cases, the value of R is less than 0.1.

III. MEASUREMENT TECHNIQUES

A gradient index fiber with a quadratic refractive index profile has focusing property. For imaging applications, the image quality is strongly dependent upon how closely the refractive index profile has been tailored to the ideal distribution. An accurate understanding of the profile distribution can help the fiber manufacturers make a GRIN fiber with an ideal profile. It can also help to evaluate the performance of different types of fiber. So, there is a need to find a fast, accurate, highly reliable and simple, but relatively low cost technique for measuring the refractive index distributions.

Several techniques have been developed to measure the index profile with consistent accuracy. They are classified as follows:

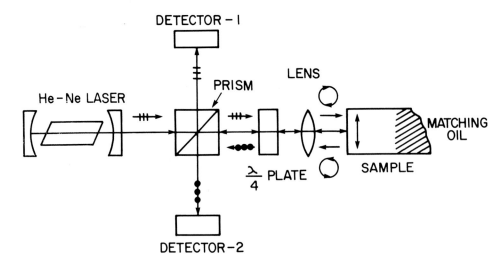

FIGURE 14.   Apparatus for reflected power measurement. (From Kokubun, Y. and Iga, K., *Trans. IECE Jpn.*, E60, 702, 1977. With permission.)

1.   Direct measurement — This earliest and simplest method directly measures the output power of an incident light passing through a fiber sample. The leakage mode and mode losses are neglected in this technique.
2.   Interferometric measurement — This is the most sensitive measurement technique. It is based on interferometric measurements either on whole fibers (nondestructive type) or on thin, polished slabs of fibers.[26] Both longitudinal[26] and transverse interference methods have been used.
3.   Refraction and focusing methods — The original concept of the refraction method was published by Chu in 1977.[27] The index profile was measured by observing rays reflected and refracted in the backward direction or the forward direction.[36] Later on, Marcuse and Presby[28] extended this idea by using a broad incoherent light for non-destructive measurement of the index profile.
4.   Aberration testing measurement[29] — The index profile is analytically related to the images in terms of aberration expressions. It does this by evaluating the focusing property of a gradient index fiber lens and its related image distortion or field curvature. However, this technique is not easily applicable at the preform stage of fibers.
5.   Nonoptical methods — Other nonoptical measurement techniques have been developed by using scanning electron microscopy,[30] X-ray microprobe analysis,[31] and chemical etching analysis.[32]

In this section, only the reflection, refraction, interferometric, and focusing methods will be discussed. More details on other methods are in References 30 to 32.

A. Reflection Method

One of the earliest techniques for measuring the index profile is the reflection method.[33] Figure 14 shows the measurement setup. A coherent He-Ne laser is used as the light source. One of the probing beams is normally incident on the surface of the testing rod. The reflected beam is collected by detector — 2. The reflectivity, R, is calculated as

$$R = \left[\frac{n(o) - n(r)}{n(o) + n(r)}\right]^2 = \left[\frac{1 - n'(r)}{1 + n'(r)}\right]^2 \tag{37}$$

where n(o) is the index of refraction at the center of the fiber rod, n(r) is an arbitrary location measured from the central axis, $n'(r) = n(r)/n(o)$, and r is the radial distance from the center. Through simple derivations, Equation 37 becomes

$$\Delta n'(r) = \left\{ \frac{[n'(o)]^2 - 1}{4} \right\} \frac{\Delta R(r)}{R(o)} \tag{38}$$

where ΔR and $\Delta n'$ are the small derivations of R and n'. The $\lambda/4$ plate and the prism are used to avoid reflected beams from other surfaces. One of the sample ends was immersed in the index matching oil to avoid the reflected beam from the surface of this end. Using manufacturer's data for n(o) allowed 4.3% of the error of the index profile to be reported by this method.[33]

B. Interferometric Method

This technique is one of the most sensitive for measuring the index profile. Interference microscopy[25] or Mach-Zehnder interferometry[34,35] are used to get interference fringe patterns from a testing fiber sample. The index profiling information is obtained from the uniform spaced distance and the shifting amount of fringe patterns.

Figure 15 shows a typical transverse interferometry (TI) that includes an interference microscopy, a video camera system, a digitizer, a programmable calculator, and a plotter. The vidicon camera with an IR-enhanced silicon target captures the interference fringe pattern from the interference microscope. The interference pattern is then digitized, and the calculator selects the necessary pixel location. The difference in refractive index between the matching oil and the measured point inside the fiber rod is given by

$$\Delta n = \frac{P\lambda}{k}$$

where k is the thickness of the sample. P is the ratio of the fringe displacement to the distance between uniformly spaced fringes generated by the matching oil. Figure 16 shows the interference patterns at $\lambda = 0.9\ \mu m$. The procedure for making the calculations is

1. Locate a desired fringe by fringe counting. For example, at a particular location on the x axis, the fringe count is four along the line AB.
2. Digitize all the pixels along AB.
3. Once the desired fringe location is found, use a parabolic fit to get about 20 data points in the neighborhood of the fringe maxima. These maxima are then taken as the value of the desired fringe center.
4. Repeat steps 1 to 3 many times by increasing the x value by 1. The averaged value of these maxima will be the fringe center on the left side of the cladding.
5. The space between uniform fringe patterns can also be found by locating the center of each fringe and then averaging over the fringes.
6. Change the location, x, to the right side of the cladding along line CD. Repeat steps 1 to 4.
7. A straight line that passes the fringe's center is determined by calculator.
8. Use this line as the reference level to find the displacement.
9. Obtain the value of P from step 5 through step 8. Then determine $\Delta n(r)$ as a function of distance r.

Generally, this setup is primarily used for a thin slab fiber sample. But, the whole fiber sample can also be used in this apparatus. The fiber rod should be immersed in the matching oil. In this case, the light should transversely illuminate the testing sample. However, a slab

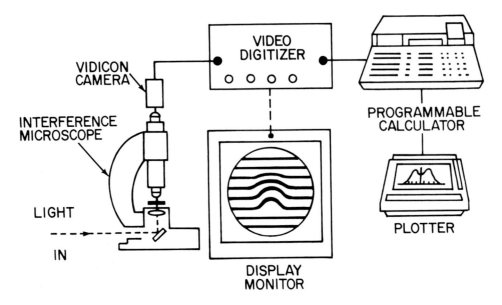

FIGURE 15. Automatic video analysis system. (From Presby, H. M., Marcuse, D., and Astle, H. W., *Appl. Opt.*, 17, 2209, 1978. With permission.)

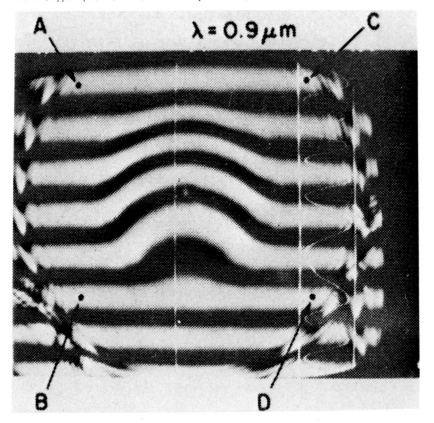

FIGURE 16. Interferogram of a graded-index fiber viewed transversely under an interference microscope at wavelength of 0.9 μm. (From Presby, H. M., Marcuse, D., and Astle, H. W., *Appl. Opt.*, 17(14), 2209, 1978. With permission.)

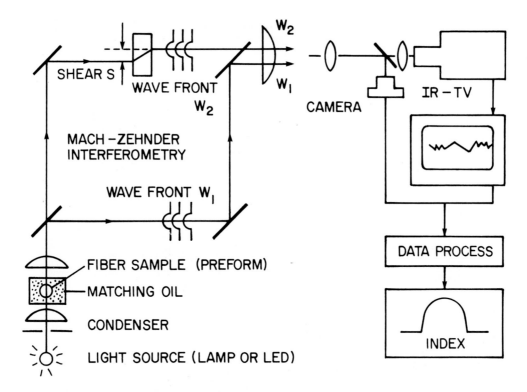

FIGURE 17. Principle of transverse differential interferometry. (From Nishizawa, K. and Momokitz, A., private communication. With permission.)

testing sample should be viewed longitudinally under an interference microscope.

The transverse interferometer (TI) is convenient, nondestructive, and accurate. It can also resolve large refractive-index fluctuations that are sharply localized. However, it has no capability for measuring the index profile of preforms. To overcome this drawback, the transverse differential interferometric[34,35] and the focusing methods[28] were developed.

Figure 17 shows the setup of a transverse differential interferometry (TDI). A Mach-Zehnder interferometer is used to divide the probing beam into two separate paths. The shearing amount, s, is introduced by a shearing prism. A sample of preform or whole fiber is immersed into index matching oil. A GaAs LED operating at $\lambda = 0.867$ μm with narrow bandwidth ($\Delta\lambda = 0.027$ μm) is used to obtain sharp fringes. Differential interference patterns are imaged by an IR-TV camera. The index profile is obtained by calculating the amounts of fringe shift.

C. Refraction Angle Method

In 1977, Chu[27] proposed a nondestructive measurement of index profile of fiber preforms. A narrow, coherent beam is shone laterally perpendicular to the optical axis. By recording the exit angles of the backscattered rays, one can get the preform index profile from the numerical integration of those exit angles. This method is also highly sensitive to the incident location of the probing beam.

A similar technique that also has high resolution was developed by Watkins in 1979.[36] It can measure the geometric and refractive index characteristics of the GRIN preform. A computer program is used to trace rays through the preform. The incident angle of the light ray is then related to its deflection in traversing the preform. The index profile is obtained by observing reflected and refracted rays in the forward direction. The technique has been widely adopted by the fiber industries.

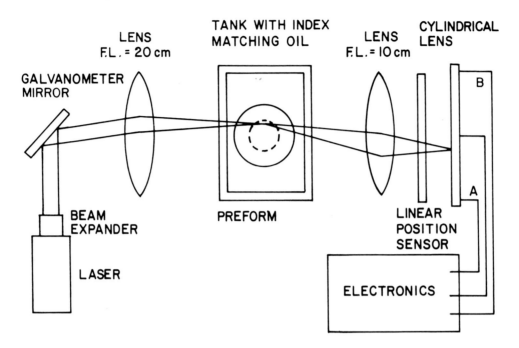

FIGURE 18. The experimental arrangement of the Watkins' method for preform measurement in index-matching oil. (From Watkins, L. S., *Appl. Opt.*, 18(13), 2214, 1979. With permission.)

The system setup of the Watkins' method is shown in Figure 18. A narrow He-Ne laser beam is focused through the preform. The galvanometer mirror controls the beam that is directed traversely across the preform. The preform should be immersed in the index matching oil to avoid the error effect caused by the noncircularity of its outside surface. The deflected beam is imaged by a lens and focused onto a linear position sensor. The beam deflection angle[36] is then calculated by measuring the centroid of the beam intensity distribution along the sensor. When the incident beam is traversely scanned across the preform, the plot of the beam position vs. the integrated deflection angle gives the index profile.

D. Focusing Method

Marcuse and Presby[28] extended the refraction angle method and developed the focusing method. This method can measure the index profile of both fibers and preforms. It is easier to use and needs less computing time. A broad band of incoherent light transversely illuminates the fibers or preforms which act as a lens. Based on geometric optics and the light power focused by the fiber lens in the near field, the index profile can be obtained from the calculation of two numerical integration equations.

1. Theoretical Background

As shown in Figure 19, this method uses the lens behavior of the fiber or preform core. A broad uniformly incident collimated beam serves as probing ray. Due to the small difference in refractive index between the center of the fiber and its periphery (or the matching oil), the paraxial ray equation can be applied in this case:[37]

$$\frac{d^2y}{dx^2} = \frac{1}{n_{mat}} \frac{\partial n}{\partial y}$$

(39)

where n_{mat} is the refractive index of the matching oil. Integrating Equation 39 and also assuming that the refractive index is a function of radial distance only, the slope of the ray which exists from the core is

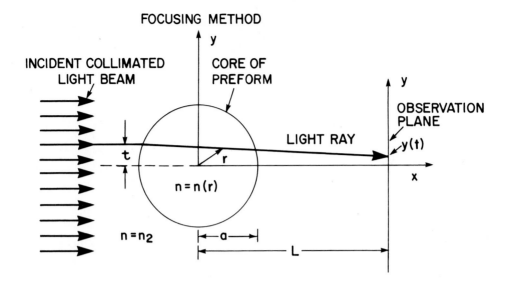

FIGURE 19. Schematic diagram of ray; trajectory needed for the theory of the focusing method. (From Marcuse, D., *Light Transmission Optics,* Van Nostrand Reinhold, New York, 1972. With permission.)

$$\frac{dy}{dx} = \frac{1}{n_{mat}} \left[- \int_a^{r_o} \frac{\partial n}{\partial r} \frac{y}{(r^2 - y^2)^{1/2}} \, dr + \int_{r_o}^a \frac{\partial n}{\partial r} \frac{y}{(r^2 - y^2)^{1/2}} \, dr \right] \qquad (40)$$

Usually, the gradient change of refractive index is small inside the fibers. One can replace $y = t$ in Equation 40. Assuming that the observation screen is located a distance, L, from the fiber's center, then

$$y(t) = t + L \frac{dy}{dx} = t + \frac{2Lt}{n_{mat}} \int_t^a \frac{\partial n}{\partial r} \frac{dr}{(t^2 - r^2)^{1/2}} \, dt \qquad (41)$$

The above integral transform can be inverted as

$$n(r) - n_{mat} = \frac{n_{mat}}{\pi L} \int_r^a \frac{t - y(t)}{(t^2 - r^2)^{1/2}} \, dt \qquad (42)$$

According to geometric optics, the power density distribution is a constant between two adjacent light rays. The amount of dy/dt is inversely proportional to the power density distribution in the near field. Since dy/dt = 1 for all rays that are outside or to the left of the fiber samples, if one can normalize the power density P to unity for $t \geqslant a$, then

$$\frac{1}{p} = \frac{dy}{dt} \qquad (43)$$

and

$$t(y) = \int_0^y P(y')dy' \qquad (44)$$

The function $y = y(t)$ can be obtained from $t = t(y)$ by sorting the number pairs, y and t.

If the power density function is measured inside the fiber samples, then Equation 42 becomes

$$n(r) - n_{mat} = \frac{1}{\pi L} \int_r^a \frac{t - y(t)}{(t^2 - r^2)^{1/2}} \, dt \qquad (45)$$

2. Measurement Procedures

Based on the theoretical analysis and the same setup as the interference method (except that the interference microscope should be changed to a regular type), the measurement procedures, briefly, are as follows:

1. Measure the power density distribution along the observation plane (see Figure 19). The location of this plane shall stay well to the left of the focal region.
2. Normalize the measured power density distribution so that P(y) is equal to unity for $y \geq a$.
3. Substitute the function P(y) into Equation 44 and obtain the function t(y) by numerical integration.
4. Invert the function t(y) by sorting out the t-y number pairs and use the function y(t) in Equation 42 to find the index profile $n(r) - n_{mat}$ by numerical integration.

This method only needs relatively simple instruments. It is not very sensitive to measurement inaccuracy and has a built-in smoothing effect to attenuate measurement errors.[36] However, the resolution of this method is low and also highly sensitive to temporal light fluctuation. Coherent light can be used in this method, but the speckle problems and interference effects also make it less convenient.

IV. FABRICATION TECHNIQUES

Several different techniques have been developed to fabricate GRIN fibers with glasses and plastics; these are: Neutron Irradiation, Chemical Vapor Deposition (CVD), Polymerization, Ion Exchange, Molecular Stuffing, and Crystal Growing. There are two important features for evaluating any of these techniques: (1) the gradient depth, and (2) the amount of change of the refractive index. It is highly recommended that readers read Chapter 2 for detailed information on optical waveguide manufacturing.

A. Neutron Irradiation

The refractive index of a boron-rich glass can be locally modified to the expected distribution profile by fast or thermal irradiation. Fast neutrons directly cause atomic displacement, which will then modify the refractive index. Thermal neutrons change the index of refraction of materials because of the secondary charged particles created in the $^{10}B(n,\alpha)^7Li$ reaction.

The original experiment[38] reported was uniform neutron irradiation in a swimming pool type Triton® reactor in France. Figure 20 is a simple schematic diagram of an irradiation device. The neutron flux (1) is directed by a collimator (2) through an appropriate flux modulation at screen (3) over the glass material (4). The refractive index with a certain expected distribution form of the glass can be modified by the neutron irradiation process.

There are some disadvantages; for example, the small gradient depth (~0.1 mm) and irradiation effect, which will certainly limit their applications in imaging systems. More research needs to be carried out in this field.

B. Chemical Vapor Deposition (CVD)

This technique has been widely used to fabricate gradient index fibers in telecommunication applications.[39-41] The basic principle of the process is to deposit glass material with a given refractive index either inside of or outside of a tube and then deposit a slightly different chemical composition as the next layer. Many layers, each with a slightly different step-

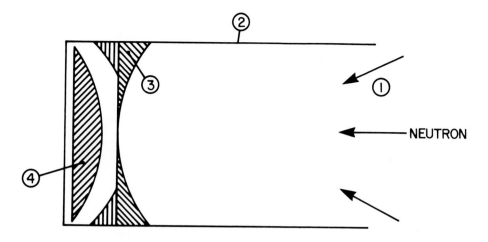

FIGURE 20.   The schematic diagram of the principle of the neutron irradiation device. Discussion on page 218. (From Sinai, P., *Appl. Opt.*, 10(1), 99, 1971. With permission.)

index, can be established by applying this process continuously. A glass rod about 2 cm in diameter is made in the original form and a fiber then drawn to a desired diameter.

Figure 21 is a schematic diagram of the CVD process.[39] Containers (1) and (2) consist of liquid constituents (3) and (4) which eventually form a layer (5). This layer is applied to a cylindrical glass starting member or rod (6) by a flame hydrolysis burner (7). Oxygen or another suitable gas with predetermined quantities and pressures is supplied to these two containers and bubbled through the liquids. The gas flow is regulated by valves (8) and (9). The flow rate is indicated by gauges (10) and (11). The liquid constituents in the containers are maintained at a desired temperature by heaters (12) and (13). When the carrier gas bubbles through container (3), the vapors of the liquid become entrained in the carrier gas and are then exhausted through pipe (14). The same thing happens for container (4). The mixed gas-vapor is fed to burner (7) and injected into flame (15). The glass soot is then formed by the hydrolyzed gas-vapor mixture. The stream (16) of soot is then directed toward starting member (6). The starting member or rod should be kept constantly rotating and translating to obtain uniform soot deposition.

The radially gradient composition desired can be deposited on the starting member rod (6) by precisely controlling the heater's temperature and predetermining the quantities and pressures and flow rate of the carrier gas to at least one of the containers. The starting member is then moved out and the remaining hollow cylinder heated. A fiber having a solid cross-sectional area with a radially varying composition can be obtained by continuously drawing and then eventually collapsing the hole of the hollow cylinder.

Generally, fused silica that has minimum light absorption is used as the base glass to form layer (5). It can be doped with different materials (such as titanium oxide, tantalum oxide, tin oxide, niobium oxide, aluminum oxide, germanium oxide, etc.) to give a radially gradient index of refraction. Because of the rather small depth of the gradient that can be achieved by this technique, its use has been limited to microoptics applications. It, certainly, has severe limitations for imaging systems that require large apertures.

A new technique called Modified Chemical Vapor Deposition (MCVD) has recently been developed at the Bell Laboratories.[41,42] The preliminary results indicated that this is inherently a clean process and enables the preforms to be fabricated much faster than by the traditional process. There are other advantages; for example, the product does not crack as often as it does in other processes and high NA fibers can be made. Another process, such as Vapor Axial Deposition (VAD), developed by NTT Laboratories in Japan, has also been used to manufacture fibers with very low losses (see Chapter 2).

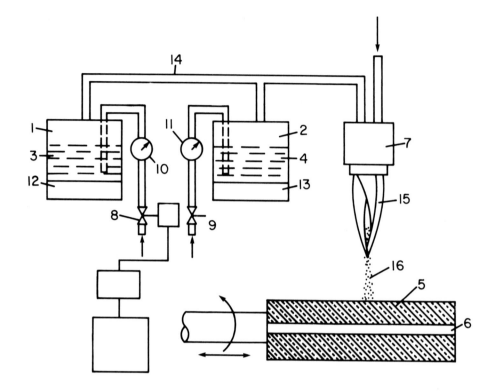

FIGURE 21.   The schematic diagram of fabricating gradient-index fibers by the CVD process. For further discussion, see page 220. (From Schultz, P., U.S. Patent 3,826,560, 1974. With permission.)

C. Polymerization Technique

As originally patented by Moore[43] in 1973, the polymerization technique is used to provide plastic lenses and lens elements having radially gradient indexes of refraction. In principle, an organic diluent that is a low molecular weight organic fluid is diffused into a substantially homogeneous, amorphous, and transparent polymetric matrix by UV irradiation or laser beam.[44] The diluent will diffuse into the polymer and then the GRIN rod can be formed.

There are two methods for making positive plastic lenses and two methods for making negative lenses. For positive lenses, the refractive index must be decreased in the outward direction. As shown in Figure 22, a diluent (1) having a refractive index lower than polymer material (2) is diffused inwardly as indicated by arrows (3). For negative lenses, one can use the same geometries for diffusion and just interchange the relative magnitude of the refractive indexes of the polymetric matrix and the diluent.

A large geometry can be made with relative ease by using this technique. However, the thermal change of gradient and the thermal effect on refractive index in plastic have rarely been measured. These kinds of thermal properties are critical for plastic lenses.

In 1973, Ohtsuka[45] successfully fabricated a light-focusing plastic rod by copolymerization of diallyl isophthalate (DAIP)-methyl methacrylate (MMA). There are two steps involved in this technique: (1) a prepolymer DAIP rod is prepared by placing the DAIP containing 3.85 wt % benzoyl peroxide (BPO) in a polypropylene tube 4 mm in diameter and 220 mm long. It is then heated at $80 \pm 0.1°C$ for about 110 min; (2) the DAIP prepolymer rod is immersed in the MMA for heat treatment at $80 \pm 0.1°C$ for about 5 min. After draining off the MMA liquid, the rod is heated at 80°C for 2 hr. During the second stage of preparation, a part of DAIP monomer will be replaced by MMA that was diffused into the rod. The copolymerization of MMA with the allylic group remaining in the rod occurs simultaneously. The radially gradient index of the MMA component is then formed.

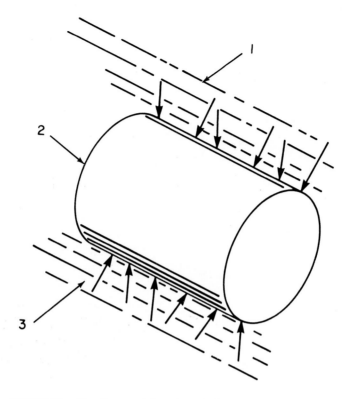

FIGURE 22. The diagram of the polymerization techniques for making a positive plastic lens. See text, page 220. (From Moore, R. S., U.S. Patent 3,718,383, 1973. With permission.)

Although the DAIP content is gradually decreased from the center to the periphery, because of the evaporation loss of MMA during the heat treatment period, it will increase again near the periphery. The unwanted layer with a high content of DAIP should be removed. The final light-focusing plastic rod is obtained by using the magnesia power to polish the cut ends.

Another technique for preparing plastic lenses is photocopolymerization. A monomer mixture of methyl methacrylate (MMA) and vinyl benzoate (VB) containing benzoyl peroxide (BPO) is polymerized in a glass tube. This tube is then exposed to UV light by rotating the tube about its axis.[47] More recently, fabrication of GRIN rods with higher clarity has been reported in which the same process is used but with the monomer pair of MMA and vinyl phenylacetate (VPA$_c$).[48]

Because GRIN rods prepared by copolymerization are infusible, they cannot be converted into a fibrous medium. However, the GRIN rods made by photocopolymerization are sufficiently fusible to be heat-drawn continuously and then converted into a fiber medium.

D. Ion Exchange

Because of the simple instruments and controls needed, this technique is so far the most popular. The alkali metal ions contained in a glass rod are replaced by larger sizes of other alkali metal ions from a fused salt bath. The index profile of the glass rod will then be modified by the different composition of ions in the glass. As a matter of fact, the ion exchange method has been widely adopted to strengthen glass material.[49,50] But the effects of modifying the refractive index of glass received little attention until 1969. Japanese scientists used this technique in late 1968 to develop the first fiber rod (the so called SELFOC® lens[51]) with a parabolic index profile for imaging applications.[52] But no information was

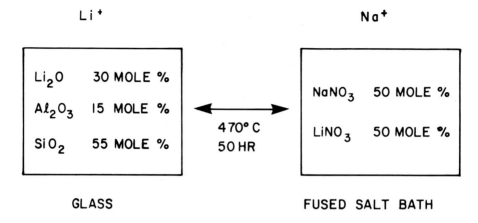

FIGURE 23.   The simple description of ion exchange method.

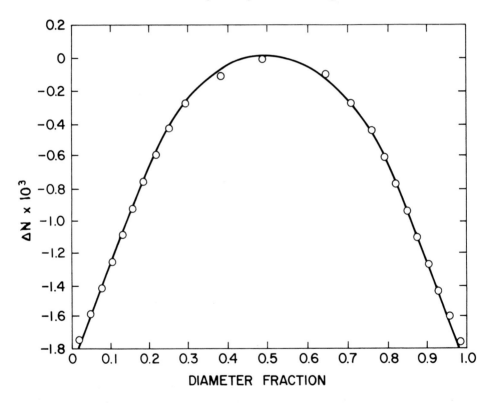

FIGURE 24.   Measured refractive index profile of an ion-exchanged rod, normalized to a maximum of zero. The refractive index at the rod center is approximately 1.539. The solid line is a parabola fitted to the exprimental points by the least-squares method. (From Pearson, A. D., French, W. G., and Rawson, E. G., *Appl. Phys. Lett.*, 15(2), 76, 1969. With permission.)

provided at that time on the production method. In 1969, the first paper describing this technique was published by Pearson et al.[53] As shown in Figure 23, the smaller Li^+ ions in a lithium alumino silicate glass were exchanged with the larger Na^+ ions in a fused salt bath at 470°C for about 50 hr. The different polarizability and the mass of these two different ions caused an index profile with parabolic form to be formed as shown in Figure 24.

Some experiments showed that a smoothly varying index profile is necessary to obtain a good image quality. At least two kinds of ions should be exchanged to get a larger difference

Table 1
REFRACTIVE INDICES OF
SILICATE GLASSES
DOPED WITH MONOVALENT
(UPPER) OR DIVALENT (LOWER)
ION OXIDE

Refractive indices of silicate glasses doped with monovalent ion oxide

	Refractive index	
Modifying oxide R_2O	SiO_2 70 mol % R_2O 30 mol %	SiO_2 60 mol % CaO 20 mol % R_2O 20 mol %
Li_2O	1.53	1.57
Na_2O	1.50	1.55
K_2O	1.51	1.55
Rb_2O	1.50	1.54
Cs_2O	1.50	1.54
Tl_2O	1.83	1.80

Refractive indices of silicate glasses doped with divalent ion oxide

	Refractive index
Modifying oxide RO	SiO_2 60 mol % RO 40 mol %
PbO	1.81
BaO	1.68
CdO	1.64
SrO	1.61
CaO	1.59
ZnO	1.58
BeO	1.54
MgO	1.51

From Kitano, I., Koizumi, K., and Matsummura, H., *Suppl. J. Jpn. Soc. Appl. Phys.*, 39, 63, 1970. With permission.

of refractive index between the center and the peripheral surface.[54] Table 1 lists the refractive indices of silicate glasses that contain an oxide of an exchangeable monovalent or divalent cations. Ion exchange between any pairs of these cations can result in a 0.01 to 0.03 of ΔN. It seems that the exchange between divalent cations can get larger ΔN; however, it takes an extremely long time. It has been reported that the ion exchange between Tl^+ and at least a kind of alkali ion was the best way to make a practical lens-like fiber rod in a reasonably short time. The ΔN of 0.04 was obtained by this method.[55] The on-axis resolving power of a fiber rod 3.14 mm long, 1 mm in diameter, and with 1.64 mm focal length is about 480 lines per millimeter. The depth of focus is usually rather poor in this case.[56]

Deviations from the idealized index profile of the GRIN causes aberrations. There are many parameters that will influence the formation of the index profile by this technique, among them being: glass composition, exchangeable ions and contents, rod diameter, bath temperature, and ion exchange time, etc. The right choice of glass composition and ion

Table 2
COMPOSITION OF
SELECTED GLASSES

Components	Type I (mol %)	Type II (mol %)
SiO_2	60	60
B_2O_3	15	20
Na_2O	15	10
Tl_2O	10	10

From Miyazawa, T., Okada, K., Kubo, T., Nishizawa, K., Kitano, I., and Iga, K., *Appl. Opt.*, 19(7), 1113, 1980. With permission.

exchange parameters can reduce the aberrations. Suggested glasses are listed in Table 2. A difference of only 5% in the amount of B_2O_3 and Na_2O can result in much different diffusion characteristics. The appropriate bath temperature is between 530 and 550°C. The glass will be deformed if the bath temperature is over 550°C. On the other hand, an extremely long time for ion exchange is needed to obtain a low aberration lens rod if the bath temperature is below 530°C. The relationships between longitudinal aberration and ion exchange time at different bath temperatures for Type I and Type II glasses are shown in Figure 25.[57] The aberrations can be reduced by increasing the ion exchange time.

Because ion-diffusion time is proportional to the square of the diameter of the prepared fiber rod, gradual variation of the ion concentration during the heating period will increase the viscosity of the glass fiber and then obstruct the ion diffusion penetration. It is quite difficult to fabricate a 5 mm diameter glass fiber by the conventional ion exchange method. In 1980, Sono et al. of Nippon Sheet Glass Co. proposed a new procedure to overcome the above drawbacks.[58] A 5 mm diameter mother rod is immersed in a potassium nitrate salt bath heated to 530°C.[58] The treating temperature is then gradually increased without deforming the glass. The treating time of this process was reduced to almost half of the time needed for fabricating a 5 mm diameter at a constant temperature.

E. Molecular Stuffing

This technique was developed by the Vitreous State Laboratory of Catholic University of America.[59-63] The method, briefly, is as follows:

1. Phase separation — An alkali-borosilicate glass consisting of 60% SiO_2, 32% B_2O_3, 4% Na_2O, and 4% K_2O is heat treated near 550°C for a suitable period of time (about 2 hr) such that two phases can be separated. One phase containing the mostly covalently bonded SiO_2 phase is called the hard phase. The other phase, called the soft phase, consists mostly of the alkali metals, boron, and transition metal impurities.
2. Leaching — The soft phase of B_2O_3 and alkali is ionic in nature. It can be dissolved out of the glass by leaching in weak acid (such as HCl), and only a glass sponge that consists of the purified hard phase will remain.
3. Stuff — After the "leaching" step, the interconnected glass porous structures are exposed to a bath and soaked in a watery dopant solution at an elevated temperature. The molecules or ions will then diffuse into the material.
4. Unstuff — After the "stuff" step, the concentration of the dopant solution is varied as a function of time in a predetermined manner. This will cause partial diffusion from the pores to obtain a desired index profile. The dopants that can be used by this process include KNO_3, $CsNO_3$, $Bi(NO_3)$, $Rb(NO_3)$, $Pb(NO_3)$, H_3BO_3, and $Sr(NO_3)_2$.[59]

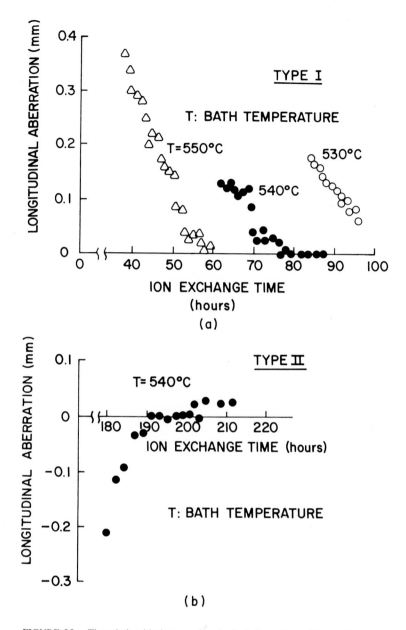

FIGURE 25. The relationship between longitudinal aberration and ion exchange time at different bath temperatures for: (a) glass type I (top); and (b) type II (bottom). (From Miyazawa, T., Okada, K., Kubo, T., Nishizawa, K., Kitano, I., and Iga, K., *Appl. Opt.*, 19(7), 1113, 1980. With permission.)

5. Crystallization — After the diffusion-controlled processes of "stuff" and "unstuff", the remaining dopant in solution is crystallized by suddenly dropping the temperature.

6. Drying, heat, and consolidation — The doped glass rod is finally dried and heated to incorporate the dopant into the glass structure and then to collapse the pores. Finally, it can be drawn into a fiber.

Figure 26 shows the refractive index profile of a GRIN rod fabricated by this method.[59] This technique offers the possibility of diffusing larger molecules or ions into the glass. Different popular profiles of refractive index can be fabricated also. Although uniformity

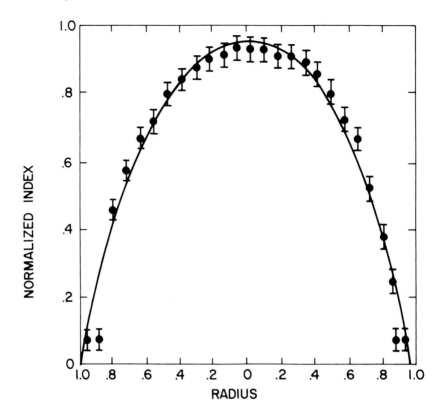

FIGURE 26. The refractive index profile for a cylindrical fiber obtained by a single step unstuffing. The solid line is the theoretically predicted profile. The index and radius are in normalized units. (From Mohr, R. K., Wilder, J. A., Macedo, P. B., and Gupta, P. K., Paper WA-1, Technical Digest on Gradient-Index Optical Imaging Systems, Rochester, New York, May 15 to 16, 1979. With permission.)

of phase separation will be a problem for the glass,[64] there is a way to control this kind of problem during the process.

F. Crystal Growing

Development of the crystal growth technique was originally stimulated by a desire to achieve an ultra low loss for IR transmissions in applying CO_2 lasers in surgery, cutting, and telecommunications applications.[65,66] Work has been done on $ZnCl_2$ glass fiber,[67] and on extruding polycrystalline fibers of thallium and silver halides.[68] The tremendous scattering loss of polycrystalline or amorphous glass fibers has caused much effort to be directed to single crystal fibers. Recently, Bell Laboratories have successfully grown single-crystal fibers with well-defined crystal orientation from AgBr.[66]

The setup of the AgBr single crystal-growing apparatus is shown in Figure 27. The liquid charge contained in a fused-quartz U tube is kept molten by a surrounding oven. The rate of liquid fed into the nozzle can be controlled by the N_2-gas pressure. The exit of the nozzle determines the cross section and the size of the growing fibers. A small oven around the nozzle tip can adjust the temperature. The size of the growing fiber can be determined by the outside diameter and the inside diameter of the nozzle. Clear and smooth fibers with diameters between 0.35 and 0.75 mm have been successfully fabricated by this process.

This kind of very low-loss transmission can also be used as infrared image dissection, and nonlinear optical devices. Using this technique, it is also possible to grow combinations of gradient index silicon and germanium for IR transmission. However, the IR materials

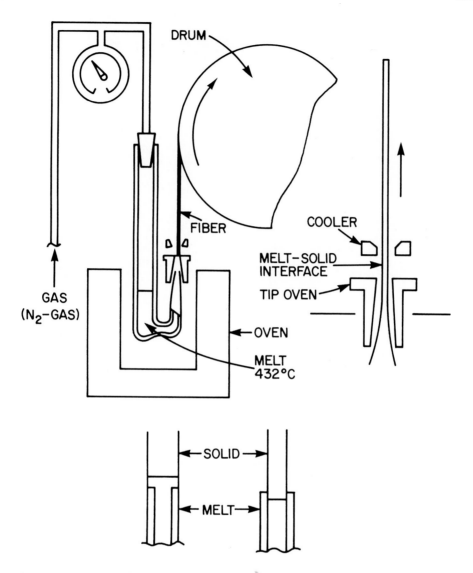

FIGURE 27. Schematic of fiber crystal-growing apparatus. (From Bridges, T. J., Hasiak, J. S., and Strand, A. R., *Opt. Lett.*, 5, 85, 1980. With permission.)

are inherently weaker than oxide glasses; thus some problems have to be solved before practical fiber lenses can be made by this process.

V. APPLICATIONS

Many applications have already been found for gradient index fibers in the optical communication field. Because of the many unique characteristics of these fibers (such as their compact size, focusing property, higher transmission efficiency, and flexibility) they have begun to be widely used in imaging applications. Each individual GRIN rod has the same imaging behavior as a single spherical lens. Figure 28 shows an object imaged by a single SELFOC® lens at different viewing angles.[69] The GRIN rod lens offers many advantages: it has a smaller diameter (<0.5 mm), it can be easily fabricated, the desired focal length can be obtained easily by choosing the right rod length (see Equation 9), it is compact with high resistance and strength, and both linear and matrix types can be made. These are advantages

FIGURE 28. The image by a single SELFOC® lens having various viewing angles: $\theta_c = 7.5°$ (right), 15° (center) and 30° (left). (From Matsushita, K. and Ikeda, K., *Proc. SPIE*, 31, 23, 1972. With permission.)

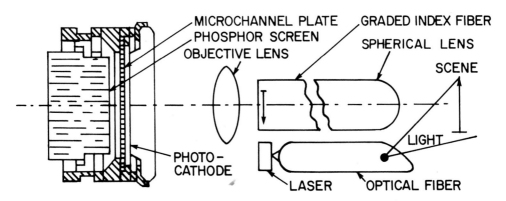

FIGURE 29. Image enhancing system. (From Csorba, I. P., *Appl. Opt.*, 19(7), 1139, 1980. With permission.)

in applications for imaging systems that need to be compact, highly flexible, and have high resolution.

The GRIN rod can be incorporated with an image intensifier as an image-enhancing system for low light level applications.[70] As indicated in Figure 29, a scene was illuminated by a GaAs laser diode. It was then imaged by the combination of a spherical lens, GRIN rod, and an objective lens at the phosphor screen. This system can convert IR radiation to visible light. A maximum luminous gain of 100,000 was obtained.

A very compact optical reader can be made by combing the GRIN fiber lens with solid state sensors and VLSI technology. OCR readers or high speed check readers are possible future applications for this system.

A GRIN linear arrayed lens is formed by packaging many individual fiber rods together. It can be used only for forming an erect image with unit magnification. Figure 30 shows

FIGURE 30.   Image obtained by a SELFOC® linear array lens. (From Matsushita, K. and Ikeda, K., *Proc. SPIE*, 31, 23, 1972. With permission.)

an image formed by a commercial linear array. Compared with a spherical lens, it has the following advantages:

1.   Uniform MTF, no cos^4 loss at the periphery: it provides uniform image quality at the edges
2.   Higher brightness: it can be used for high speed copiers/duplicators
3.   Short conjugate length: it provides a compact optical system
4.   High strength and resistance

Figure 31 shows two different imaging systems for a copier/duplicator using a spherical lens and a GRIN linear array fiber lens. A much shorter conjugate length is needed in the fiber optics case. Figures 32 and 33 also show an interesting comparison of the typical differences between a Xerox 3100 copier lens and a standard SELFOC® lens.[71] Only unit magnification can be used by a fiber linear array lens, therefore, the copier/duplicator that requires the reduction mode still needs a conventional spherical lens. Figure 34 also shows a nonimpact printer application that uses a linear fiber array lens, in this case a line-scanning CRT, to obtain an image on photosensitive material.

So far, all our discussions and descriptions of applications have been limited to fibers that have the parabolic form of gradient-index distribution. The reasonably good depth of focus (~1.3 mm) makes this kind of fiber optics useful as an imaging system for plain paper copiers. However, for low cost and low volume applications, one can also use the step-index fiber as an optics information transmitting medium. Because of the inherently very poor depth of focus, direct contact is needed to improve image quality. In such copiers, coated paper should be used.

Pontarelli, as early as 1970, demonstrated[72] a facsimile scanner using fiber optics. It provided uniform light distribution and constant resolution across the entire scan line. Subsequently, 3M announced their "283 copier" that uses a fiber optics bundle with step-index as a transmitting medium to transfer the optical information from the document to the photosensitive area.[73] Most recently, Bell Laboratories[74] used optical fibers for automatic scanning and digitizing engineering drawings with 250 samples per inch (spi) resolution. A shown in Figure 35, the fiber optics illuminator and receiver should be in direct contact with the document. A light source illuminates the circular end of the fiber guide and then illuminates a narrow strip across the document. The fiber receiver will pick up the reflecting

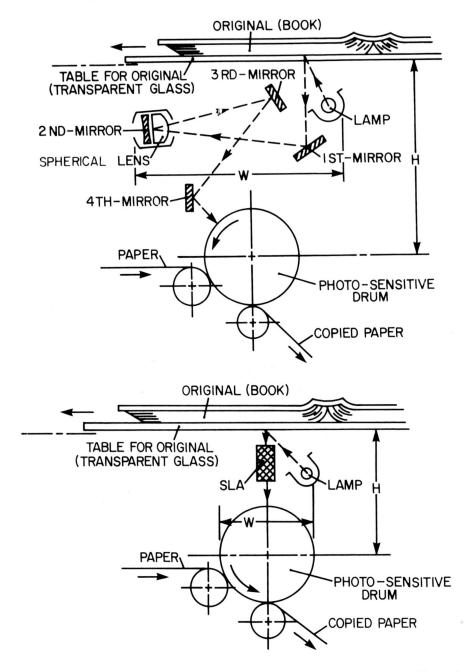

FIGURE 31. The imaging system of a copier/duplicator by way of conventional spherical lens and GRIN linear arrayed lens (bottom). (From Nippon Sheet Glass Co. Ltd., Data Sheet, Osaka, 1979. With permission.)

optics and then transmit the image to the other end as a matrix grid shape. The advantages of high flexibility of the fiber bundle and uniform flux distribution at the illuminated end are clearly shown in this design. The only drawback to using fiber bundles in this design is "broken fibers". However, this problem can be overcome by a carefully controlled process during the assembly.

As one can see, image scanners using fiber optics with their compact size, high resolution, and high speed can have new potential applications for computer-aided design and engineering.

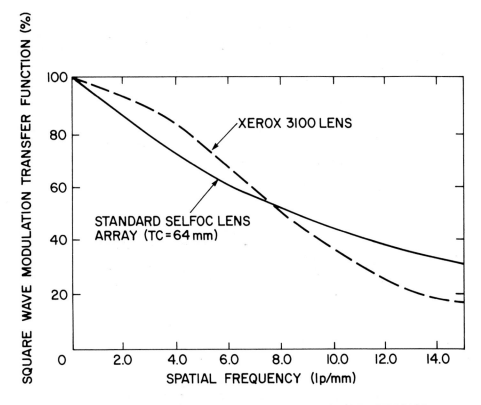

FIGURE 32. The average square wave MTF of a typical standard gradient index SELFOC® lens array and a typical (nongradient) Xerox 3100 copier lens as a function of spatial frequency. (From Rees, J. D., SELFOC® '81, Exploration of SELFOC® Technology, NSG America, Clark, N. J., 1981. With permission.)

OPTICAL SYSTEM	XEROX 3100	HYPOTHETICAL SELFOC SYSTEM
TOTAL CONJUGATE	760 mm	64 mm
TYPICAL LENS BEST FOCUS SQUARE WAVE MTF @ 6.4 lp/mm	~63%	~60%
TYPICAL LENS DEPTH OF FOCUS @ 40% MODULATION AND 6.4 lp/mm	~1.0 mm	~1.0 mm
RELATIVE PHOTORECEPTOR EXPOSURE	1.00	0.83

FIGURE 33. A summary comparison of a Xerox 3100 copier conventional optical system and a hypothetical gradient index SELFOC® lens array optical system. (From Rees, J. D., SELFOC® '81, Exploration of SELFOC® Technology, NSG America, Clark, N.J., 1981. With permission.)

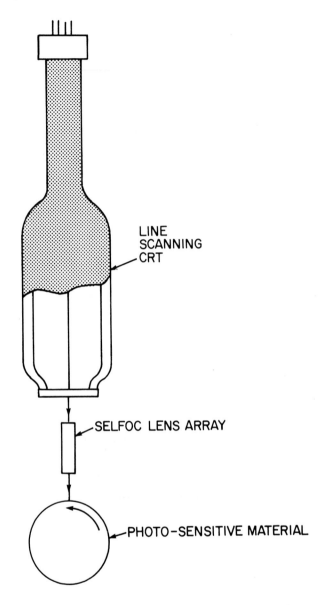

FIGURE 34.   A line-scanning CRT printer using a SELFOC® lens array for imaging purposes. (From Rees, J. D., SELFOC® '81, Exploration of SELFOC® Technology, NSG America, Clark, N.J., 1981. With permission.)

VI. FUTURE READING

   It would be impossible to discuss all the related topics in detail, but the references of this chapter are a useful bibliography of this subject. Readers wanting a deeper understanding of this should not only peruse the chapters in this book, but also the following recommended books and journals:

Marchand, E. W., *Gradient Index Optics,* Academic Press, New York, 1978.
Marcuse, D., *Principles of Optical Fiber Measurements,* Academic Press, New York, 1981.
Special issue on Gradient Index Optics for Imaging, *Applied Optics,* April 1, 1980.

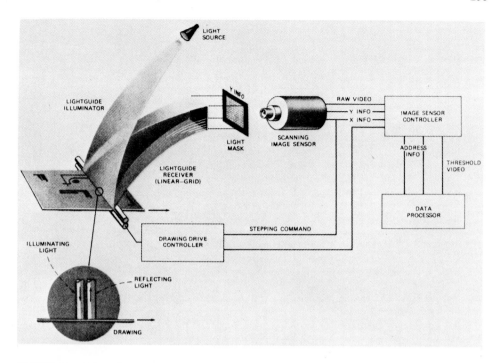

FIGURE 35.   Schematic representation of an optical fiber scanning digitizer. (From Pferd, W. and Stocker, G., *Bell Sys. Tech. J.*, 60, 523, 1981. Copyright 1981, American Telephone and Telegraph Company. Reprinted by permission.)

Special issue on GRIN2, *Appl. Opt.*, March 15, 1982.
Digest of the topical meeting on Gradient-Index Optical Imaging Systems, Honolulu, May 4 to 5, 1981.
Brown, S. J. S., An introduction to gradient index optics, *Marconi Rev.*, 1st Quarter, 1981.
Okoshi, T., *Optical Fibers,* Academic Press, New York, 1982.

ACKNOWLEDGMENTS

I wish to thank all those who responded to my request for information on their publications. Many invaluable discussions were held with N. Seachman and J. Rees. Thanks to Dr. K. L. Yip and Dr. H. T. Shang for carefully reading the original manuscript and their many suggestions. The enthusiastic encouragement from W. Haas is also appreciated. Finally, thanks are also due to Dr. P. Lavallee for his constant support when the author was with the Xerox Corporation.

PROBLEMS

1.   Describe the similarity between conventional singlet spherical lens and GRIN fiber lens.
2.   Derive Equation 6.
3.   Derive Equation 14.
4.   The use of chemical vapor deposition (CVD) in fabricating fibers causes small sinusoidal perturbations in refractive index down the length of fiber. It will affect image quality. Describe this phenomena quantitatively.
5.   Why can the fiber linear array lens be used in the case of unit magnification only?

REFERENCES

1. **Maxwell, J. C.,** *The Scientific Papers of James Clerk Maxwell,* Nives, W. D., Ed., Dover Publ., New York, 1965.
2. **Wood, R. W.,** *Physical Optics,* MacMillian, New York, 1905.
3. **Marchand, E. W.,** Gradient index imaging optics today, The Topic Meeting on Gradient-Index Optical Imaging Systems, Honolulu, May 4 to 5, 1981.
4. **Yariv, A.,** *Introduction To Optical Electronics,* Holt, Rinehart & Winston, New York, 1971.
5. **Kapron, F. P.,** Geometrical optics of parabolic index-gradient cylindrical lenses, *J. Opt. Soc. Am.,* 60, 1433, 1970.
6. **Marcuse, D.,** *Light Transmission Optics,* Van Nostrand Reinhold, New York, 1972, 86.
7. **Suematsu, Y. and Iga, K.,** Mode conversion in light beam waveguide, *J. IECE Jpn.,* 49, 1645, 1966.
8. **Kornhauser, E. T. and Yaghjian, A. D.,** Modal solution of a point source in a strongly focusing medium, *Radio Sci.,* 2, 299, 1967.
9. **Kawakami, S. and Nishizawa, J.,** An optical waveguide with the optimum distribution of the refractive index with reference to waveform distortion, *IEEE Trans. Microwave Theory Tech.,* MTT-16, 814, 1968.
10. **Rawson, E. G., Herriott, D. R., and McKenna, J.,** Analysis of refractive index distributions in cylindrical, graded-index glass rods (GRIN rods) used as image relays, *Appl. Phys.,* 9, 753, 1970.
11. **Rees, J. D. and Lama, W.,** Some radiometric properties of gradient-index fiber lenses, *Appl. Opt.,* 19, 1065, 1980.
12. **Matsushita, K. and Toyama, M.,** Unevenness of illuminance caused by gradient-index fiber arrays, *Appl. Opt.,* 19, 1070, 1980.
13. **Kawazu, M. and Ogura, Y.,** Application of gradient-index fiber arrays for copying machine, *Appl. Opt.,* 19(7), 1105, 1980.
14. **Levi, L. and Austing, R.,** Tables of the modulation transfer function of a defocused perfect lens, *Appl. Opt.,* 7, 967, 1968.
15. **Seachman, N.,** personal communication, 1981.
16. **Moore, D. T. and Sands, P. J.,** Third-order aberration of inhomogeneous lens with cylindrical index distribution, *J. Opt. Soc. Am.,* 61, 1195, 1971.
17. **Sands, P. J.,** Third-order aberrations of inhomogeneous lenses, *J. Opt. Soc. Am.,* 60, 1436, 1970.
18. **Gupta, A., Thyagarajan, K., Goyal, I. C., and Ghatak, A. K.,** Theory of fifth-order aberration of graded-index media, *J. Opt. Soc. Am.,* 66, 1320, 1976.
19. **Streifer, W. and Paxton, K. B.,** Analytic solution of ray equations in cylindrically inhomogeneous guiding medium 1: meridional rays, *Appl. Opt.,* 10, 769, 1971.
20. **Tomlinson, W. J.,** Aberrations of GRIN-rod lenses in multimode optical fiber devices, *Appl. Opt.,* 19(7), 1117, 1980.
21. **Nishizawa, K.,** Chromatic aberration of the Selfoc® lens as an imaging system, *Appl. Opt.,* 19(7), 1052, 1980.
22. Nippon Sheet Glass Co. Ltd., Data Sheet, Osaka, Japan, 1979.
23. **Iga, K.,** Theory for gradient-index imaging, *Appl. Opt.,* 19, 1039, 1980.
24. **Marcatili, E.,** Modes in a sequences of thick astigmatic lens-like focusers, *Bell Syst. Tech. J.,* 43, 2887, 1964.
25. **Presby, H. M., Marcuse, D., and Astle, H. W.,** Automatic refractive-index profiling of optical fibers, *Appl. Opt.,* 17(14), 2209, 1978.
26. **Martin, W. E.,** Refractive index profile measurements of diffused optical waveguides, *Appl. Opt.,* 13, 2112, 1974.
27. **Chu, P. L.,** Nondestructive measurement of index profile of an optical fibre preform, *Electron. Lett.,* 13, 736, 1977.
28. **Marcuse, D. and Presby, H. M.,** Focusing method for nondestructive measurement of optical fiber index profiles, *Appl. Opt.,* 18(1), 14, 1979.
29. **Yamamoto, N. and Iga, K.,** Evaluation of gradient-index rod lenses by imaging, *Appl. Opt.,* 19(7), 1101, 1980.
30. **Burrus, C. A. and Standley, R. D.,** Viewing refractive-index profiles and small scale inhomogeneities in glass optical fibers: some techniques, *Appl. Opt.,* 13, 2365, 1974.
31. **McKinley, T. D., Heimich, K. F. J., and Whittry, D. B., Eds.,** *The Electro Microprobe,* John Wiley & Sons, New York, 1966.
32. **Presby, H. M., Standley, R. D., MacChesney, J. B., and O'Connor, P. B.,** Material structure of germanium-doped optical fibers and preforms, *Bell Syst. Tech. J.,* 54, 1681, 1975.
33. **Ikeda, M., Tateda, M., and Yoshikiyo, H.,** Reflective index profile of a graded index fiber: measurement by a reflection method, *Appl. Opt.,* 14, 814, 1975.
34. **Nishizawa, K. and Momokita, A.,** Index profiling measurement of SELFOC® lens by computer-controlled transverse differential interferometry, private communication, 1981.

35. **Kokubun, Y. and Iga, K.,** Precise measurement of the refractive index profile of optical fibers by a nondestructive interference method, *Trans. IECE Jpn.,* E60, 702, 1977.

36. **Watkins, L. S.,** Laser beam refraction traversely through a graded-index preform to determine refractive index ratio and gradient profile, *Appl. Opt.,* 18(13), 2214, 1979.

37. **Marcuse, D.,** *Light Transmission Optics,* Van Nostrand Reihold, New York, 1972.

38. **Sinai, P.,** Correction of optical aberrations by neutron irradiation, *Appl. Opt.,* 10(1), 99, 1971.

39. **Schultz, P.,** Method of Forming Focusing Fiber Waveguide, U.S. Patent 3,826,560, 1974.

40. **Keck, D. B. and Olshanasky, R.,** Optical Waveguide Having Optimal Index Gradient, U.S. Patent 3,904,268, 1975.

41. **MacChesney, J. B., O'Connor, P. B., and Pearson, A. D.,** Optical Fiber Process, U.S. Patent 4,191,545, 1980.

42. **MacChesney,** Materials and processes for preform fabrication — modified chemical vapor deposition and plasma chemical vapor deposition, *Proc. IEEE,* 68, 1181, 1980.

43. **Moore, R. S.,** Plastic Optical Element Having Refractive Index Gradient, U.S. Patent 3,718,383, 1973.

44. **Swainson, W. K.,** Method, Medium and Apparatus for Producing Three-Dimensional Figure Product, U.S. Patent 4,041,476, 1977.

45. **Ohtsuka, Y.,** Light-focusing plastic rod prepared from diallyl isophthalate-methyl methacrylate copolymerization, *Appl. Phys. Lett.,* 33, 247, 1973.

46. **Ohtsuka, Y. and Nakamoto, I.,** Light-focusing plastic rod prepared by photocopolymerization of methacrylic esters with vinyl benzoates, *Appl. Phys. Lett.,* 29, 559, 1976.

47. **Ohtsuka, Y., Koike, Y., and Yamazaki, H.,** Studies on the light-focusing plastic rod. 6: The photocopolymer rod of methylsmethacrylate with vinyl benzoate, *Appl. Opt.,* 20, 280, 1981.

48. **Koike, Y., Kimoto, Y., and Ohtsuka, Y.,** A light-focusing plastic rod of methyl methacrylate — vinyl phenylacetate, paper TU B1-1, Technical Digest of Gradient-Index Optical Imaging Systems, Honolulu, May 4 to 5, 1981.

49. **Zijstra, A. L. and Burggraaf, A. J.,** Fracture phenomena and strength properties of chemically and physically strengthened glass, *J. Non-Crystalline Solids,* 1, 49, 1968.

50. **Doremus, R. H.,** *Ion Exchange — A series of Advances,* Vol. 2, Marinsky, J. A., Ed., Marcel Dekker, New York, 1969, chap. 1.

51. SELFOC® is a registered trade name of the Nippon Sheet Glass Co., Ltd. of Osaka, Japan.

52. **Uchida, T., Furukawa, M., Kitano, I., Koizumi, K., and Matsummura, H.,** A light-focussing fiber guide, IEEE Conf. Laser Engnr. Appl., Washington, D.C., 1969, 331.

53. **Pearson, A. D., French, W. G., and Rawson, E. G.,** Preparation of a light focusing glass rod by on-exchange techniques, *Appl. Phys. Lett.,* 15(2), 76, 1969.

54. **French, W. G. and Pearson, A. D.,** Refractive index changes produced in glass by ion exchange, *Am. Ceram. Soc. Bull.,* 49, 974, 1970.

55. **Kitano, I., Koizumi, K., and Matsummura, H.,** A light-focusing fiber guide prepared by ion-exchange techniques, *Suppl. J. Jpn. Soc. Appl. Phys.,* 39, 63, 1970.

56. **Kita, H., Kitano, I., Uchida, T., and Furukawa, M.,** Light-focusing glass fibers and rods, *J. Am. Ceram. Soc.,* 54, 321, 1971.

57. **Miyazawa, T., Okada, K., Kubo, T., Nishizawa, K., Kitano, I., and Iga, K.,** Aberration improvement of Sefoc® Lenses, *Appl. Opt.,* 19(7), 1113, 1980.

58. **Sono, K., Nishizawa, K., Yamamoto, N., Momokita, A., Kobayashi, M., and Kitano, I.,** Fabrication of gradient index lenses whose diameter is large, Electro-Optical Technical Symp. SPIE, Huntsville, Ala., September 29 to October 2, 1980.

59. **Mohr, R. K., Wilder, J. A., Macedo, P. B., and Gupta, P. K.,** Graded index lenses by the molecular stuffing process, paper WA-1, Technical Digest on Gradient-Index Optical Imaging Systems, Rochester, New York, May 15 to 16, 1979.

60. **Mohr, R. K., Gupta, P. K., Drexhage, M. G., Hojaji, H., Simmons, J. H., and Macedo, P. B.,** Strengthening of optical fibers by molecular stuffing, in *Fiber Optics,* Bendow, B. and Mitra, S. S., Eds., Plenum Press, New York, 1979, 143.

61. **Simmons, J. H., Mohr, R. K., Tran, D. C., Macedo, P. B., and Litovitz, T. A.,** Optical properties of waveguide made by a porous glass process, *Appl. Opt.,* 18, 2732, 1979.

62. **Macedo, P. B. and Litovitz, T. A.,** Method for Producing an Impregnated Waveguide, U.S. Patent 4,110,093, 1978.

63. **Macedo, P. B. and Litovitz, T. A.,** Method of Producing a Glass Article Having a Graded Refractive Index Profile of a Parabolic Nature, unpublished work.

64. **Moore, D. T.,** Gradient-index optics: a review, *Appl. Opt.,* 19, 1035, 1980.

65. **Mimura, Y., Okamura, Y., Komazawa, Y., and Ota, C.,** Growth of fiber crystals for infrared optical waveguides, *Jpn. J. Appl. Phys.,* 19, L269, 1980.

66. **Bridges, T. J., Hasiak, J. S., and Strand, A. R.,** Single-crystal AgBr infrared optical fibers, *Opt. Lett.,* 5, 85, 1980.

67. **Pinnow, D. A., Gentile, A. L., Standlee, A. G., Timper, A. J., and Hobrock, L. M.,** Polycrystalline fiber optical waveguides for infrared transmission, *Appl. Phys. Lett.,* 33, 28, 1978.
68. **VanUitert, L. G. and Wemple, S. H.,** $ZnCl_2$-glass: a potential ultra low-loss optical fiber material, *Appl. Phys. Lett.,* 33, 57, 1978.
69. **Matsushita, K. and Ikeda, K.,** Newly developed glass devices for image transmission, *Proc. SPIE,* 31, 23, 1972.
70. **Csorba, I. P.,** Image enhancement for light focusing graded index optical fibers, *Appl. Opt.,* 19(7), 1139, 1980.
71. **Rees, J. D.,** SELFOC® '81, Exploration of SELFOC® Technology, NSG America, Clark, N.J., 1981.
72. **Pontarelli, D. A.,** Continuous facsimile scanner employing fiber optics, *Proc. SPIE,* 21, 59, 1970.
73. **Bleeker, L. A. and Owen, R.,** Fiber Optical Element Imaging and Illumination Assembly, U.S. Patent, 4,194,827, 1980.
74. **Pferd, W. and Stocker, G.,** Optical fibers for scanning digitizer, *Bell Sys. Tech. J.,* 60, 523, 1981.

INDEX

N